Dinil Gajbhiye
Swarup Deb
Prasad Gorase

Low Cost Automation And Effective Material Handling Systems

D1744023

Dinil Gajbhiye
Swarup Deb
Prasad Gorase

Low Cost Automation And Effective Material Handling Systems

LAP LAMBERT Academic Publishing

Impressum/Imprint (nur für Deutschland/only for Germany)
Bibliografische Information der Deutschen Nationalbibliothek: Die Deutsche Nationalbibliothek verzeichnet diese Publikation in der Deutschen Nationalbibliografie; detaillierte bibliografische Daten sind im Internet über http://dnb.d-nb.de abrufbar.
Alle in diesem Buch genannten Marken und Produktnamen unterliegen warenzeichen-, marken- oder patentrechtlichem Schutz bzw. sind Warenzeichen oder eingetragene Warenzeichen der jeweiligen Inhaber. Die Wiedergabe von Marken, Produktnamen, Gebrauchsnamen, Handelsnamen, Warenbezeichnungen u.s.w. in diesem Werk berechtigt auch ohne besondere Kennzeichnung nicht zu der Annahme, dass solche Namen im Sinne der Warenzeichen- und Markenschutzgesetzgebung als frei zu betrachten wären und daher von jedermann benutzt werden dürften.

Coverbild: www.ingimage.com

Verlag: LAP LAMBERT Academic Publishing GmbH & Co. KG
Dudweiler Landstr. 99, 66123 Saarbrücken, Deutschland
Telefon +49 681 3720-310, Telefax +49 681 3720-3109
Email: info@lap-publishing.com

Herstellung in Deutschland:
Schaltungsdienst Lange o.H.G., Berlin
Books on Demand GmbH, Norderstedt
Reha GmbH, Saarbrücken
Amazon Distribution GmbH, Leipzig
ISBN: 978-3-8454-0825-5

Imprint (only for USA, GB)
Bibliographic information published by the Deutsche Nationalbibliothek: The Deutsche Nationalbibliothek lists this publication in the Deutsche Nationalbibliografie; detailed bibliographic data are available in the Internet at http://dnb.d-nb.de.
Any brand names and product names mentioned in this book are subject to trademark, brand or patent protection and are trademarks or registered trademarks of their respective holders. The use of brand names, product names, common names, trade names, product descriptions etc. even without a particular marking in this works is in no way to be construed to mean that such names may be regarded as unrestricted in respect of trademark and brand protection legislation and could thus be used by anyone.

Cover image: www.ingimage.com

Publisher: LAP LAMBERT Academic Publishing GmbH & Co. KG
Dudweiler Landstr. 99, 66123 Saarbrücken, Germany
Phone +49 681 3720-310, Fax +49 681 3720-3109
Email: info@lap-publishing.com

Printed in the U.S.A.
Printed in the U.K. by (see last page)
ISBN: 978-3-8454-0825-5

Low Cost Automation And Effective Material Handling Systems

At

BAJAJ AUTO LTD., Akurdi, Pune.

By

Dinil Gajbhiye

Prasad Gorase

Swarup Deb

ACKNOWLEDGEMENT

A successful project work is the result of teamwork, which contains not only the people who put their logic, but also who guide them.

It gives us immense pleasure in expressing our sincere gratitude and appreciation to all those who have been instrumental in the successful completion of the project.

The most important person whose efforts have been monumental in the completion of this project is our project guide Mr. BHATTU for his kind co-operation during the project. He was very generous with his advice, views and ideas and always ensured that we were on the right track.

We wish to express our sincere gratitude to Dr. S R Kajale for his kind guidance, valuable suggestions and encouragement.

We wish to express out profound sense of gratitude to our guides at BAJAJ Auto., Pune, Mr. S. R. DESHPANDE (DEPT. MANAGER AL. SHOP) and Mr. GAWAS (ASST. MANAGER AL. SHOP) for the keen interest they showed in our project.

Dinil Gajbhiye
Prasad Gorase
Swarup Deb

CONTENTS

COMPANY PROFILE

1. INTRODUCTION:

The Bajaj Group is amongst the top 10 business houses in India. Its footprint stretches over a wide range of industries, spanning automobiles (two-wheelers and three-wheelers), home appliances, lighting, iron and steel, insurance, travel and finance. The group's flagship company, Bajaj Auto, is ranked as the world's fourth largest two- and three- wheeler manufacturer and the Bajaj brand is well-known in over a dozen countries in Europe, Latin America, the US and Asia.

Founded in 1926, at the height of India's movement for independence from the British, the group has an illustrious history. The integrity, dedication, resourcefulness and determination to succeed which are characteristic of the group today, are often traced back to its birth during those days of relentless devotion to a common cause.

The present Chairman and Managing Director of the group, Rahul Bajaj, took charge of the business in 1965. Under his leadership, the turnover of the Bajaj Auto the flagship company has gone up from Rs.72 million to Rs.46.16 billion (USD 936 million), its product portfolio has expanded from one to and the brand has found a global market. He is one of India's most distinguished business leaders and internationally respected for his business acumen and entrepreneurial spirit.

Figure 1 Places where BAJAJ Markets its products

The highlights of its international dominance are:

Distribution network covers 50 countries

1,56,007 units exported in 2003-04

Dominant presence in Sri Lanka, Mexico, Bangladesh, Columbia, Guatemala, Peru, Egypt, Iran and Indonesia

Largest exporter of three-wheelers; over 65.797 units exported in 2003-04

All products customized as per market needs

66 per cent growth in total exports in 2003-04

In countries where Bajaj perceives a good market potential, it seeks a tie up with one of the major industrial establishments, which would be in a position to invest in the project and which would also entail manufacturing activities apart from marketing, distribution and after sales services through a well-established nation-wide network.

2. ACCOMPLISHMENTS:

The first and the only automobile exporter from India to clock exports of 100,000 units in one financial year (in 2003-04)

Consistent winner of the Engineering Export Promotion Council (EEPC) Awards for the last decade

Largest manufacturer exporter in terms of the product range offered

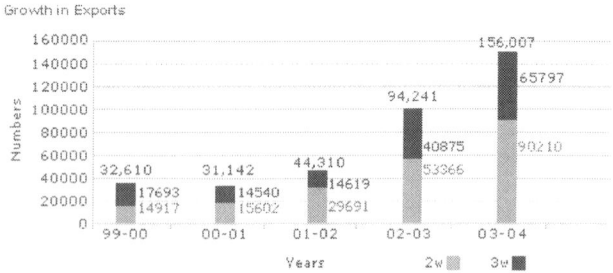

Figure 2 Graph of growth in export

3. Manufacturing Departments

All the machining operations are done in these departments:

Steel Machining Shop

Aluminum Machining Shop

6

The components are then sent for assembly or other processes to the following shops:

Paint shop

Chassis and other parts are painted.

Engine assembly

Main engine is assembled.

Vehicle assembly (final assembly)

The final assembly of the automobile is done.

Quality control services

Quality of different parts that are manufactured is verified.

Manufacturing engineering dept.

Production engineering dept.

The existing models are modified according to the demand from the market.

R&D department

New products are developed.

4. Bajaj Products:

Figure 3 Products of BAJAJ

Low Cost Automation

1. Introduction

The rapid growth of technology in the last two decades has been characterized by full automation. As only the affluent and developed countries can afford to adopt full automation, a large productivity gap has been created. This gap can be bridged through low cost automation. Although the less developed countries are in need of mass produced goods for their large population, it is not practical for them to adopt full automation because of limitations in capital resources and technical skills. But they must have maximize the return on their limited resources. Their desire to increase productivity can effectively accomplished through the adoption of low cost automation by their characteristic small and medium size companies.

The concept and principles of low cost automation are basically the same as those of full automation except that the former builds improvement around existing equipment and machine systems rather than replacing whole system with automated processes. The Design of low cost automation may be considered a technical task, but it is a task simple enough for a person who is technically inclined, yet without any formal technical education.

2. Automation

Automation is the use of energy of non-living systems to control and carry out a process operation without direct human intervention. The objective of automation is to make the best use of available resources, material, money and machines. With automation, human labour acquires new qualitative measures, becoming more complex and meaningful.

The essential need for Automation can be traced fundamentally to the population distribution of the world. Over the last century and more, the average life expectancy of human beings has been on the rise thanks to scientific development in every sphere of life. This, along with a variety of other factors has led to a population distribution in a manner such that the number of 'passive' people, who are not playing any role in the production process and are solely consuming goods, is steadily rising. In addition, the number of 'active' people, that is people who are working in any producing activity cannot support this increasingly burdensome segment of the population. Thus, there is a requirement for developments that could help man support more people by producing more goods, without sacrificing additional time and effort which in other words is Automation.

3. What is low cost automation?

Low cost automation is a means of simplifying production jobs and / or providing speed and accuracy for the work element involved. In broader Technical terms, low cost automation can be defined as a system that consists of relating co-coordinating integrating machines, mechanisms, devices and controls to the extent that partial or complete processing of production inputs is accomplished automatically, with or without the assistance of human beings.

4. Why lost cost automation?

Technology has made available to us devices that can see better, hear better sense better and measure better without getting fatigued than human operators. These devices and mechanism are more reliable, more powerful, more precise, more consistent and fully predictable. Unlike human workers, they do not complain of overwork, deter productivity or require a regular increase in cost. There are thousands of operations performed by human which could be performed more accurately, more efficiently and at a lower cost by automatic means, using mechanical and electrical devices that can be obtained easily.

Why not take full advantage of all possible improvements and adopt full automation? There are several reasons why manufacturing concerns hesitate to plunge into fully - automated processes. Among them are: -

a) Limitations in financial resources. Automation equipment is generally custom built and therefore may be expensive.

b) Demand for the product may be unpredictable or of insufficient volume to justify the applications of mass production techniques.

c) To obtain full benefits from automation requires more than simply superimposing automatic mechanisms on existing production operations. It requires a thorough study of the product and the process which is time consuming and expensive.

d) Labourers general resistance to radical changes and their resistance in accepting anything that will pose a threat to their work even for reasons of increased productivity.

e) The economic reasoning of management, who prefer to spend funds on old machines rather than buy new and more efficient equipment.

By selecting areas where potential benefits are greatest, the rewards of additional investment will be substantial. Some of the reasons why low cost automation is desirable and appropriate can be summarized as below.

a) The existence of a market for products where large producers cannot compete with small-scale manufacturers (i.e. demand is not large enough to justify the smallest fully automatic plant.).

b) The demand for the products exceeds the productivity of small- scale producer but is not large enough to justify plant expansion (i.e. adding, another production line).

c) Ease of operation directly increases productivity. Plant output can be increased upto 20 per cent by implementing low cost automation improvements.

d) Ease of operation results directly in satisfied workers. Improved working condition, like a safer working place and minimum physical fatigue contribute significantly to the workers' satisfaction.

e) Ease of operation results in better treatment of tools. Simplification of production process directly improves the utilization of tools and equipment.

f) Change is less abrupt, the process of adoption less destructive, and resistance correspondingly weaker.

5. Characteristics of low cost automation.

Several plants are known to have rejuvenated more effectively by automating as many phases of operation as possible than by replacing old equipment and machines of human activities. Where previously the operation of machines was paced by workers now machines pace the workers.

Application of low cost automation does not require the replacement of existing machines. Certain activities in the process can be made automatic by installing simple devices and mechanisms in the system or the machines itself.

It is seen that by introducing low cost automation

a) The labour can be deployed for the best advantage.

b) Reduces human effort and error

c) Reduces the wastage of material.

d) Increases output.

e) Operator can control several machines.

f) Ensures consistency and accuracy.

6. Area of application

Since low cost automation consists of replacing human attributes and functions in the production process, then the concept is applicable to activities requiring the human functions of sensing, control and manual work. In the context of manufacturing these activities can be classified as receiving information, decision making, remembering. Acting, exerting force and controlling.

Application of low cost automation to plant operations may range from single lines of automatic equipment, such as Presses which are fed and cycled continuously with automatic handling to complete manufacturing facilities where components are fed, sorted, oriented assembled and tested automatically. Low cost automation finds its biggest application in job lot, small volume and / or intermittent processes.

Industries in which low cost automation has been applied, include primary metals, fabricated products, machinery, telephone and telegraphy, transportation equipment, food, tobacco, textile, timber and wood, furniture, paper printing, chemicals petroleum, rubber, leather, glass and clay, mining, merchandising and distribution.

7. Workers and low cost Automation

A worker in an industry is subjected to various types of stresses. They may be occurring because of work itself, and the environment. These stresses can be due to temperature, humidity, radiant heat of the surrounding and air speed (Climatic factors) and other factors (non climatic) such as rate of work, the amount and type of clothing, physical fitness, age sex, etc. All these together influence the worker and develop physical fatigue in him. Apart from this, the worker is subjected to mental fatigue due to his relation in work - place and home.

In a mass producing industry where productivity, quality with consistency and reliability are very important, it is found that the average productivity level of work performed by manual operation is very low. It is because the worker soon finds himself exhausted, and in some cases even though the machine has substantial capacity to produce, the worker cannot cope with it. In an autocratic environment, the worker may work for longer duration but after few days or months, he may develop chronic fatigue. Then is a stage when he finds himself tired in spite of full night sleep. This ultimately results in sickness and absenteeism.

The modern outlook is that human being should do more human job (i.e. intellectual work) than doing inhuman or primitive type of work.

Low cost automation is the most effective tool in relieving human beings from fatigue, his productivity has gone up by 100 %. The areas at which the fatigue can be eliminated are.

    a) Work performed by hand and legs.

    b) Repetitive work.

c) Work requiring force.

d) Inaccessible areas.

8. Criteria for Selecting LCA

In searching for areas of potential improvement, where low cost automation is applicable, it will be preferable to perform this search in stages to conserve time and effort. One criterion to determine which areas should be studied and improved is return on investment - i.e what improvement will yield the highest returns Another criteria could be urgency i.e. what improvements must be accomplished first? Tables of priority can, therefore, be established from a rough estimate of the magnitude of the benefits to be derived, and the urgency of the needed improvements.

The first stage in the search for improvement (i.e. Cost reduction) can be made in the following areas:-

a) The product - its value analysis

b) Unit production costs -analysis of cost break down.

e) Plant operation-

eg   1) Hours of operation per week

        2) Plant / equipment utilization

        3) Rejected finished products.

        4) Manufacturing cycle time.

        5) Type of Process

        6) Product demand.

        7) Bottleneck

        8) Accidents

Application of Industrial Engineering Techniques.

The second stage in search for improvement requires a closer look or a more detailed analysis of the various tasks involved in the production process. This search can be accomplished by the application of such industrial engineering techniques, as:-

a) Methods of improvement

b) Arrangement of the work place.

a) Methods of improvement

Several studies have been conducted regarding ways and means of improving work performance. It has been found that as much as 50 per cent of work done manually in factories, offices and homes can be improved, reducing the corresponding effort expanded and the cost incurred .

b) Arrangement of the workplace

The layout of the working place greatly influence the performance of the tasks of the workers. The same is true when machines or devices are utilized since awkward and complex movements are likely to complicate installation and operation of these machines.

9. Standard devices for low cost automation

Basic functions performed by low cost automation devices

The second stage in identifying areas of potential improvement is better accomplished with an understanding of the better functions performed by low cost automation devices. These functions which are also useful in identifying the work elements of the tasks in the production process , are

    a) Receiving information

    b) Acting

    c) Exerting forces

    d) Controlling

    e) Remembering

    f) Decision-making

a) Receiving Information

This task is performed by human being either using one or a combination of sensory devices eyes for seeing, noses for smelling, tongues for tasting, ears for hearing, skin for sensing temperature, pressure or vibration. For these sensory organs the following devices are available as substitutes

    1) Photoelectric cells for the eyes

    2) Smoke detectors for the nose

    3) pH meters for the tongue

    4) Ultrasonic pickup for the ears

    5) Thermisters, bellows and Vibrometers for the skin Load cells

However, it must be realized that relative to their human counterparts, these mechanical devices have limited capabilities.

b) Acting

This task deals with motion characterized by direction. There are three major types of motion, liner (reversible or unidirectional) Circular (Clockwise, Counter- clockwise or

14

both) and a combination of both linear and circular. These motions can be performed by electric motor, linkages, cams, and gear trains and by combination of these combined linear and circular motion can be achieved.

c) Exerting forces

The application of force can be grouped into three categories pulling pushing and twisting. A pneumatic or a hydraulic cylinder can exert, through piston action, either pulling or pushing force while the shaft of an electric motor can provide twisting force directly. These mechanical devices that supply force replace the function of the human muscles.

d) Controlling

The elements of control can be defines as the regulations of action and the exertion of force. In the human body, the brain nerve muscle network performs this function. Control is vital for automating task. Some of the technical systems which can furnish certain types of control are Cams electrical relays pneumatic and hydraulic valves, transistor tubes and other electronic gadgets.

e) Remembering (Memory)

This is a function of human brain. A technical system that can replace this function would be a Computer. However, there are certain mechanical devices that can be made to remember in limited manner. This function often overlaps with the decision making function.

f) Decision-making

This is another function of human brain. In the production process, a man's experiences and knowledge can be built in through programs or predetermined instructions. Mechanical decision making is simple, limited and repetitive. Programming machines to perform decision-making tasks is generally dependent upon feed back information, which the machines receive.

Standard Devices

The work performed by standard low cost automation devices can also be classified into two types, Viz.

a) Holding and / or positioning devices:

15

Jigs and fixtures

b) Control and / or force- acting devices:

I.   Pneumatic

II.   Hydraulic

III   Electrical and Electronics

Control Devices

Pneumatic Devices: e.g. air power cylinder, air motions, control valves and various safety accessories for the operation of machine tools.

Area of application:

a) Fixtures for clamping work- pieces prior to machining.

b) Devices for feeding work - pieces into the process during machining cycle.

c) Devices for ejecting work- pieces upon completion of machining cycle.

d) Devices for shifting control levers

e) Devices for tripping valve controls for automatic sequencing

f) Devices for driving machine spindles by the use of forming press.

h) Devices for moving machine elements for retracting and positioning work and other similar applications.

1) Air Power cylinders are normally non -rotating, reciprocating piston types and are available in single and double acting cylinders. Air cylinders can considerably reduces the required time of loading, feeding, sizing gauging and segregating. Following factors should be considered while selecting air cylinders.

Force Air Pressure Piston Area

Force = Air Pressure x Piston Area.

2) Air operated boosters: With the help of air operated boosters, low-pressure air supply can be efficiently used for delivering high pressure oil in oil; hydraulic application. These are much more cheaper than hydraulic reservoir and pump usually required for hydraulic operations.

3) Pneumatic control valves: These are of various types e.g. Manual operated, Pilot operated Solenoid operated or combination of any of these. Solenoid and Pilot operated valves can be remotely controlled.

Hydraulic Devices

The primary disadvantages of an air cylinder is its inability to give precise movements to the piston because of the compressibility to use hydraulic cylinders.

Hydraulic system consists of the following components:

a) Pressure Pumps

b) Cylinders, Pistons, Boosters and Motors

c) Pressure Control Valves, Sequence Valves, Loading and loading Valve, Switch Valve, Relief Valves etc.

Hydraulic Pumps: They are of three types

a) The Gear Pump

b) Vane Pump

c) Piston type Pump

Horsepower rating can be calculated as below.

Hp = 0.000583 x Fluid volume (gpm) x Fluid Pressure (pai)

Hydraulic Cylinders: They are similar to air cylinder except that in place of air , oil is used.

Other hydraulic items are: -

Time Delay Switches: They give a delay from 0.25 to 60 Seconds.

Hydraulic Pressure Boosters: They are used to boost system pressure up to 300 kg. / cm^2

Electrical component and Devices

Electrical devices are most useful for monitoring and controlling operations from a place remote from the equipment being operated. Some of the most commonly used electrical devices are as follows:

Capacitors Resistors, Toggle Switches, Solenoids, Electrical heating elements, Electric motors, Electrically operated timers.

Mechanisms linkages and Connectors

Although the applications of low cost automation is based on the fact that existing equipment will not be replaced by specialized equipment (whenever possible) and that improvements should be built around existing or standard equipment, it often becomes necessary to convert certain motions to other types. Conversion of motion and transmission of force are best accomplished by means of mechanisms, linkages. These together form kinetic linkages. These are highly important for automated system, computers, instruments, controls and in general high speed machinery.

There are various types of mechanisms some are as follows.

a) Slider Crank Mechanism

b) Cam Mechanism

c) Four bar Mechanism

d) Quick Return Mechanism

e) Toggle Link

F) Straight Line Mechanism

g) Elliptical Trammel

h) Parallel Motion Mechanism

i) Gear Trains.

Flexible Connectors: Flexible Connectors such as flat Belts V Belts and Roller Chains are useful for transmitting power where control distances are too great for gearing and they are also used as conveyors and hoists. Examples of these are-

a) Flat Belts

b) V Belts

c) Timing Belts

d) Roller Chains

e) Variable Speed Drive

f) Universal Joints.

10.Design of Low Cost Automation System

The concept of automation / automatic control is primarily characterized by the capacity for self-regulation. A pre selected input variable or function is controlled which gives the desired output. The function of control can therefore be defines as the regulation of output quantities of a system (e.g. position, speed, temperature, force etc.) in accordance with an input command with attributes related to the desired output

There are basically two types of self-regulating systems, viz.:

a) Open - loop Control Systems

b) Closed loop Control Systems.

a) Open- loop Control Systems: In the open loop control system, the elements are connected in a linear sequence, which is open at both ends. The input and output is of definite nature. For example, in an automatic fire alarm, as soon as higher temperature is detected due to fire, this will trigger the fire protection system i.e. operate the ire alarm and start the water pump. But this will continue even if the fire extinguishes, To stop it manual intervention is necessary. There is no automatic feed back.

The advantage of open loop system is simplicity and low cost.

b) Closed loop control system: Closed-loop control systems, sometimes called feed back control are based upon the monitoring of the output, and the comparison of the output with the input command signal. The input and output influence each other. The control cycle starts with an input command signal that triggers a characteristics output response. Information is then fed back to the input to serve as a basis for correcting or adjusting deviations from the desired output. For example the use of an automatic thermostat to maintain a certain room temperature in air conditioning installation is an example of a closed loop control system. The thermostat continuously monitors room temperature

The advantage of closed-loop control system are highly accurate, quick response, flexibility and less dependence on operating conditions.

Choice of control system

The choice of control system to adapt is normally influenced by the Production process.

For batch-type or intermittent production process both open control and closed control systems are applicable. The major consideration is the cost of the system, and should be justified by return or Investment. The closed system is more complex, more costly and requires higher maintenance cost.

In case of continuous process industry, closed loop system is more appropriate as the production process requires information feed-back system for each phase of operation.

Available means of control

Control systems are implemented by any one or the combination of the following means:

    a) Mechanical

    b) Pneumatic

    c) Hydraulic

d) Electrical

e) Electronics

Selection of low cost automation areas

a) Any operation which requires muscle force or power. the average man capacity is limited to about 30 Kg. force and 0.1 KV power , and it is Unreasonable to demand as such as this all day long.

b) Any repetitive sequences of operations Men can find a working rhythm but may also lose it.

c) Any co-operation which must be done quickly e.g. pushing an emergency buffer. Our reaction time is about 0.23 Second , even when we are expecting the signal.

d) Judgment as to Process is finished. Our five Senses are inferior to instruments.

e) Judgment of quality , i.e. inspection, instruments are frequently better.

Other factors to be considered for designing low cost automation

a) Distance between the work ands the control station.

b) Working space available

c) Ambient condition of the working place

d) Cost and source of energy

e) Reliability of control required.

f) Skill required of the operator and maintenance crew.

g) Effect on and relationship with other operations in the plant.

11. Examples of LCA :

Low cost automation has implemented in various areas, some of them are as follows a) Handling

b)   Clamping on fixtures

c)   Feeding

d)   Assembly operation

e)   Linear indexing/Locating

f)   Rotary indexing

g)   Door control

h)   Bulk feeding

12. Limitation of LCA

a)   Only low cost allowed

b)   Most of the time manual labour is not eliminated

20

c)   Modification of LCA are not easy to conceive

d)   Aesthetic sense is not always possible

e)   In handling areas , where the job is located, fixed or moved , the feel to choose right device is very important

f)   Sometimes, it is slower   than   manual   work   i.e.   even though   the fatigue is saved ,cycle time has increased

g)   If the improvement are not exploited immediately, the additional expenditure may go waste

13.Conclusion

In a developing country like India, where industrialization is taking place rapidly , one of the biggest problem that we are facing is the maximum utilization of available resources in a competitive market . One of the objectives of 7th five year plan is productivity improvement. It is seen that most of our industries are not even operating at 50% productivity of an equivalent Japanese company with same manpower. We cannot always blame the workers for their low output, at the same time we cannot imagine of using highly automated machines operating without man, this will be too costly proposal at the same time may not be advisable because of lack of know –how of high technology and its availability in country. The path that can be conveniently chosen is that of maximizing the efficiency of available sources of production. This can be conveniently done with the help of low cost automation The equipments and ideas used are most convenient to adopt in any small to large scale industry . As an investment benefit analysis ,the attached graph will tell the extent of benefits.

        The industries   where   one   can   easily   adopt   LCA   with   immediate   returns   on investment are :

a)   Automobiles

b)   Textiles

c)   Agriculture

d)   Machinery Manufacturing

e)   Small scale industries

f)   Offices

g)   Hospitals

h)   Laboratories

i)   Services, etc.

Low cost automation is a technique which can be used in stages ,
and it yields advantages after each stage of automation as shown in the graph attached.

All improvements need not be introduced in one stage , as sometimes it becomes
very difficult to rapid changes .Japan is one of the countries which have derived maximum
benefits of low cost automation. It is seem that in Japan alse high cost automation has
come gradually that too only after the demand has gone to very high level .

JOB AUTO-LOADER

1. EXISTING METHODOLOGY

The generalized layout of manufacturing lines at BAL has been shown in the figure.

The storage space for components to be machined is at the start of manufacturing line. Generally plastic bins are used to carry the jobs (6-10 jobs per bin depending upon the job size). The bins are placed in flow racks that are arranged as shown in figure.

Powered conveyors run in front o the machines on which the jobs are transported in a single line one by one.

The task of loading the bins (carrying jobs) on the rack from the raw material storage is done by contractors. Loading of the racks is done 4-5 times per shift in which about 16-18 bins are loaded at a time. The operator on the first machine empties the bin by removing two jobs at a time and stores 6-8 jobs on the conveyor. The empty bin is then manually placed on the upper stage of the rack.

2. LIMITATIONS OF THE EXISTING METHODOLOGY

1. The operator on the machine 1 has to attend to more than 1 M/ C at a time. Unloading the jobs from the bin and placing the empty bins on the second stage of rack, thus becomes a very tedious job for the operator.

2. Manual handling of the jobs from the bin was the first hurdle for the future scope for automating the line.

3. Moreover the time involved in unloading the jobs from bins should have actually been used for some more productive job.

3. PROJECT OBJECTIVE

Seeing the future scope of line automation, there was a need of an auto unloader, which would load the jobs from bins to the conveyor line.

The objectives of the project were:

- To design and fabricate a model of auto-loaded that needs minimum alteration in the existing layout and equipments.
- To make the automation suitable for implementing on all the lines.
- To eliminate manual handling while loading the jobs.
- To deploy labour into productive jobs.
- To impose ideas for future implementation of the line.

24

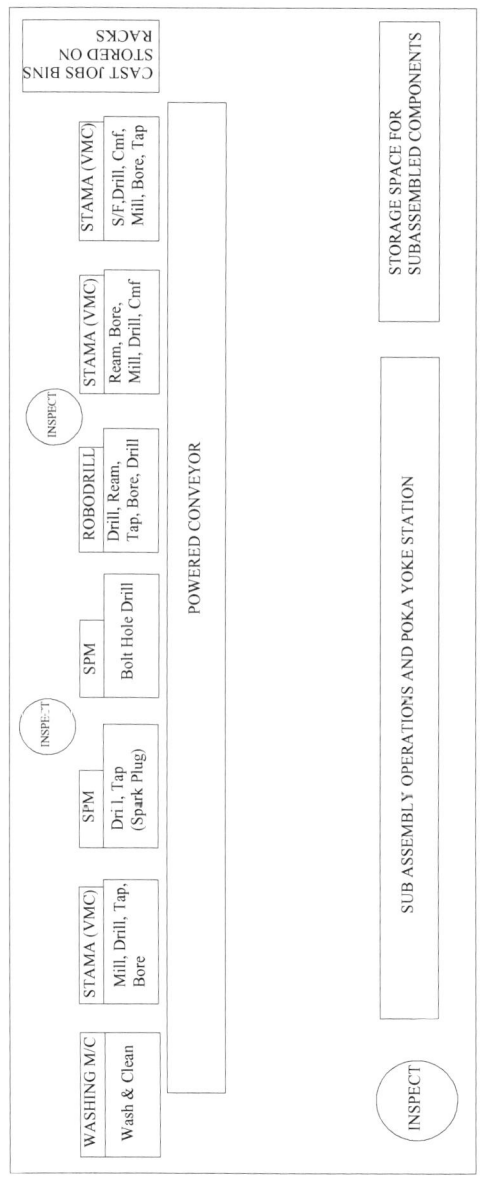

Figure 4 Layout of 3W4S Cylinder Head line BAL

4. CONCEPT AND WORKING

A basic concept of the autoloader was developed is as described below:

The auto-loader assembly consists of basically three units:

i. The flow rack, with bins containing jobs to be placed on the upper stage and empty bins on the lower stage.

ii. Job unloader is a box with rollers on the lower face and double flaps on the upper face. The box is of the size to just fit in a bin. A counter-weight is attaches to the job unloader unit. The job unloader unit and counter weight is pivoted.

iii. Powerless/ gravity tray type conveyor.

The autoloader will work as explained below:

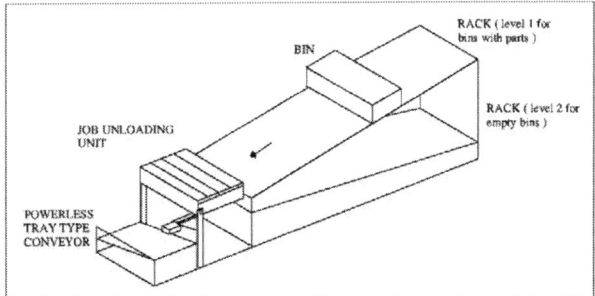

Figure 5: Concept Step 1

- STEP1: As shown in the figure, the bins containing jobs in the upper stage of the racks flow into the job unloader unit due to gravity.

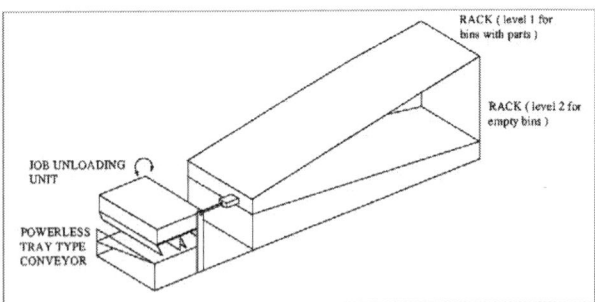

Figure 6: Concept Step 2

- STEP2: The job unloader unit rotates through 180° about an axis passing through its edge. The flaps open after the unit is rotated through 180°, thus creating passage for the jobs to fall out from bins.

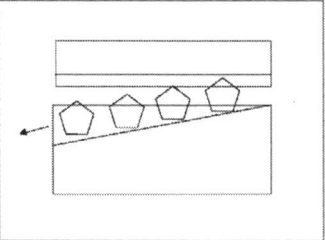

Figure 7: Concept Step 3

- STEP 3: The jobs extract from the bin are directed towards the operator using power-less gravity conveyor as shown in the figure. The operator then picks up the job to load on machine. Once all the parts from the tray are emptied the tray returns to its original position (below the job loading unit) due to the counter-weight attached to it.

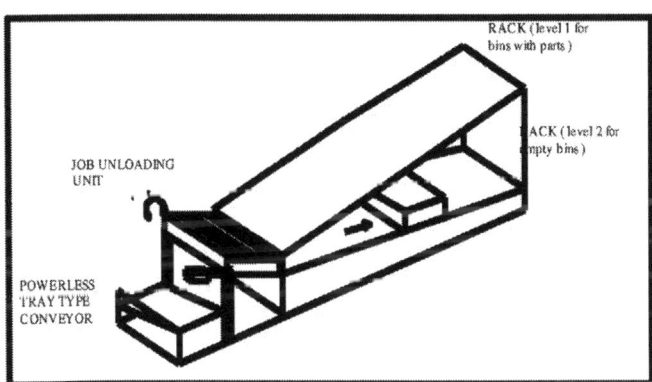

Figure 8: Concept Step 4

- STEP 4: The job loading unit rotates back to align with the lower stage of the rack, thus allowing the empty bin to escape. So, it is clear that auto-loading of job is possible using the above concept.

To ensure that automation performs consistently and is fool-proofed, it was necessary to co-ordinate each and every event taking place.

It was decided to use pneumatic double acting piston cylinders, motors, and pneumatic valves for achieving the purpose.

As the piling up of bins was not possible on the rack in this automation there was a need to increase the stages of the rack and convert it into a continuous flow rack. This would ensure a

27

greater stock of bins on the rack. To convert the rack to continuous flow rack a counter weight operated tray was attached between the top two stages.

5. FABRICATION AND DESIGN

The actual design and fabrication of the auto unloader had to go hand in hand as many unknown design parameters could only be obtained after some parts were fabricated.

The job unloading unit was chosen for the fabrication first.

Figure 9: Photo of JUU (Fabricated)

A conveyor having the shown dimensions was cut and channels were welded on it to make a box open from just one side for a bin to fit in.Next, a pair of hinged flaps was welded on the top side of frame of the job unloading unit.

Figure 10: JUU

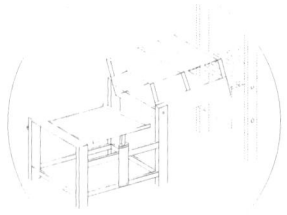

Figure 11: Assembly of JUU

Brackets were suitably welded on the job unloading unit to facilitate the pivoting of the same about its bottom edge.

The job unloading unit was weighed without and with bin full of parts. The data was used to design the counter weight and torque of the pneumatic motor.

- Selection of pneumatic motor and design of counter-weight.:-

  i. Weight of empty job unloading unit =   18.7 kg

  ii. Weight of bin (with jobs)   = $1.5 + 1.2 \times 8$ kg

  iii. 3W4S cylinder head jobs   =   11.1 kg

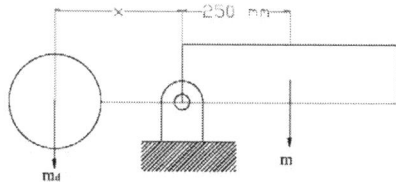

Figure 12: counter weight on JUU

Selecting pneumatic motor of 20 Nm torque $\approx$ 30 kg

Case1. Consider that job unloading unit has bin

$$30 \times 0.25 \times 9.81 - md \times x \times 9.81$$

$$= 20 \text{ Nm}$$

Assuming x = 0.25m

$$md = 21.8 \text{ kg}$$

Case2. Finding the torque needed to rotate the job unloading unit when empty,

$$18.7 \times 0.25 \times 9.81 + 21.8 \times 0.25 \times 9.81$$

$$= T'$$

$$= 8.085 \text{ Nm}$$

The torque required in both cases has large difference. So as to reduce the maximum torque required m0 was increased to 24 kg Torque for case1 is

$$30 \times 0.25 \times 9.81 - 24 \times 0.25 \times 9.81 = T1$$

$$T1 = 14.715 \text{ Nm}$$

For case 2

$$(24 \times 0.25 \times 9.81) - (18.7 \times 0.25 \times 9.81)$$

$$= 13.217 \text{ Nm} = T1$$

Considering for FOS 1.25, the pneumatic motor was selected having a torque = 18Nm of the following specifications:-

Semi rotary vane driver

29

Diameter = 16mm   swivel angle = 0°- 270° (Adjustable)

Torque = 18 Nm   Double acting

Piston Dia. = 40mm

The counter-weight was designed to be 24kg. Next, the dimensions of the rack were decided as shown.

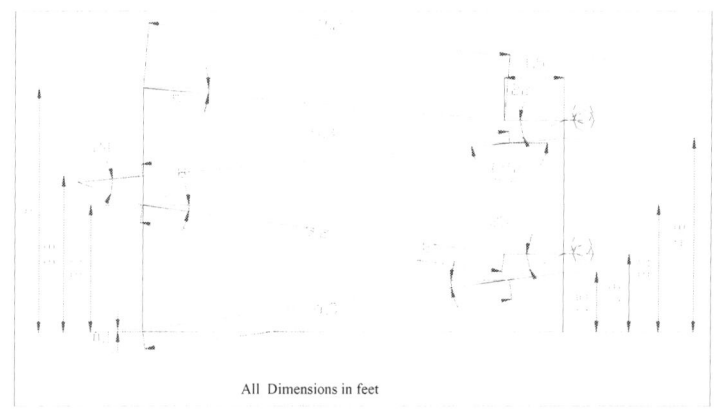

All Dimensions in feet

Figure 13: Dimension s of modified rack

The dimensions were decided such that it could stock 14 bins with parts ready for auto loading.

The productivity of 3W4S line was = 580 parts/ shift

i.e., no of bins per shift = 72.5≈73

Hence the rack needs to be loaded (72.5/ 14) ≈ 5 times a shift, where earlier only 4 reloads were sufficient as number of bins being loaded at a time was 18. To fabricate the rack, an existing was modified that had dimensions close to our requirement. The existing rack was a non flow type rack having empty bins carrying rollers on the upper stage and filled bins carrying rollers in the lower stage.

The following modifications were carried out as per the figure shown.

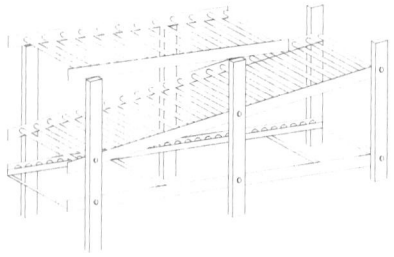

Figure 14: Modified Rack

30

i. The roller stages were interchanged.

ii. Both the stages were shifted down to accommodate another stage at the top.

iii. An additional empty bin carrying roller stage was added at the bottom most level.

iv. The rack was constructed into a continuous flow rack by attaching trays with counter-weight, to allow the transfer of bins between different levels on gravity.

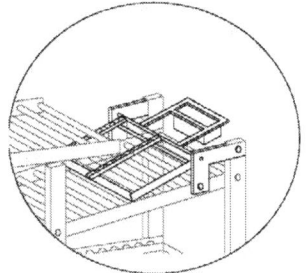

Figure 15: Gravity Tray of Continuous Flow Rack

• Design for the counter-weight trays

Two separate designs were needed for the transfer of a) heavy bins, b) empty bins.

a) Heavy bins:

Mass of one job= 1.2kg   mass of bin = 1.5kg

Mass of tray was measured to be 5.6kg

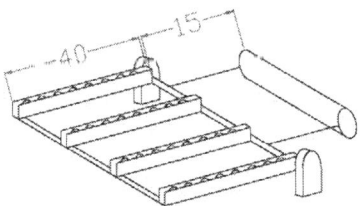

Figure 16: Gravity Tray for Heavy Bin

Thus, total mass = mass of tray + mass of a bin with parts

= 16.7kg

This weight acts at (20+5) cm from the hinge point. Thus

$16.7 \times 25 = W \times 15$ (assuming counter wt. acting at 15 cm from the hinge)

Thus, W= 27.83kg

For quick bin transfer and slow return of the tray, the counter wt. was decreased to 20 kg.

31

Figure 17: Photo of Gravity Tray

b) Empty bin:

Wt. of the tray was measured to be 2.3kg, hence taking moment

$2.3 \times 25 = W = 3.233$kg ≈ 4kg ... (assuming counter wt. acting at 15cm from the hinge.)

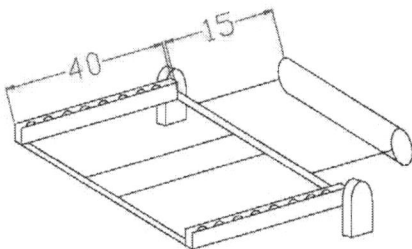

Figure 18: Gravity Tray for Empty Bin

The powerless conveyor which was to be used to carry the parts extracted from bin to carry the operator on the principle explained in section 1 of "Effective Material Handling systems".

- Design of Gravity Tray:

  Wt. of the tray was measures to be 2.5 kg

  Distance to be traveled by the tray 2.5 kg

  Taking no revolution of pulley n= 0

  R= Radius of bigger pulley

  r = Radius of smaller pulley

  h= height traveled by the counter weight.

  n (2πr) = 250

  Thus, R=4cm

  n (2πr) = h =100 cm

  Assuming h=100cm

  R= 1.6cm

32

$\cot\theta = 1/\mu \times [Pw/(g \times \{2m1 + Pw/g\})]$

$\cot\theta = 1/0.06 \times [8 \times 1.2/(2 \times 2.5 + 8 \times 1.2)]$

Thus, $\theta = 5.2°$.

Height of the gravity conveyor near the operator as per the ergonomics principles was taken to be 934mm

Thus calculating the height required at the job unloader unit

$$h1 = 934 + 250\sin5° = 115.889$$

$$= 115cm$$

$$\approx 3.8 \text{ feet}$$

which is available at the job unloader unit.

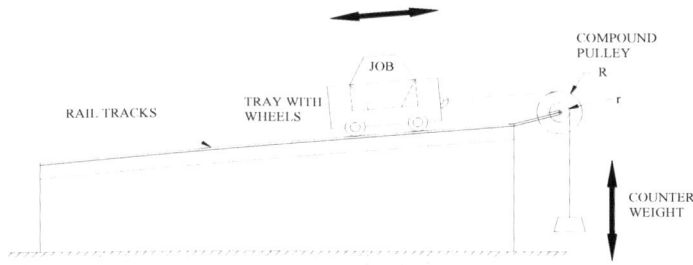

Figure 19: Concept of Powerless tray

6. CIRCUIT DESIGN:

Refer to the detailed figure showing the working.

Piston P1, stops the bins on level 1 from sliding to the counter-weight operated tray.

Piston P2 locks the counter weight operated tray.

Piston P3 stops the in on level 2 from entering the job unloading unit.

Piston P4 operates the flaps of the job unloading unit.

Piston P5 supports the job unloading unit while it aligns with the stage 2.

Piston P6 stops the job unloading unit at level 2.

A pneumatic motor is used to rotate job unloading unit.

A proper sequence of operating the pneumatic components was established to ensure the continuous, consistent and fool proof operation of the Autoloader as follows.

Event No.	Motor	P1	P2	P3	P4	P5	P6
Initial	1	1	0	1	1	1	1
1.	2	1	0	1	1	0	0
2.	2	1	0	1	0		0
3.	3	1	0	1	0	0	0
4.	1	1	0	1	1	0	1
5.	1	0	1	0	1	1	1
6.	1	1	0	1	1	1	1

NOTATIONS USED:

(Motor Positions)

1. Job unloading unit as aligned with level2.
2. Job unloading unit turns to make 180° with horizontal.
3. Job unloading unit reverses to align with the level3.

(Piston)

1. Piston at TDC
2. Piston at BDC

Initial event: - Bins with the parts inside the job unloading unit. The logic developed above was further used to design the pneumatic circuit for the automation.

Four proximity switches and one limit switch was used to generate signals for the actuation of the pistons and motors.

PS1 proximity sensor mounted at the end of the job unloading unit, it senses the presence of bin in the JUU.

PS2 proximity sensor mounted on the frame of the JUU which senses the reaching of the JUU to the station 2.

PS3 proximity sensor mounted near the frame of the JUU which senses the falling of the jobs.

PS4 proximity sensor mounted at the tip of the piston cylinder P6, it senses the stopping of the JUU at the station 1.

LS limit switch that is mounted in the third level of the rack, it generates electrical signal when the empty bin passes by striking it.

4/2 valves are used for operating all the pneumatic devices. An electrical switch is provided near the operator that disconnects the limit switch from the circuit which stops the system only after completing all its operations and the JUU reaches to position 3.

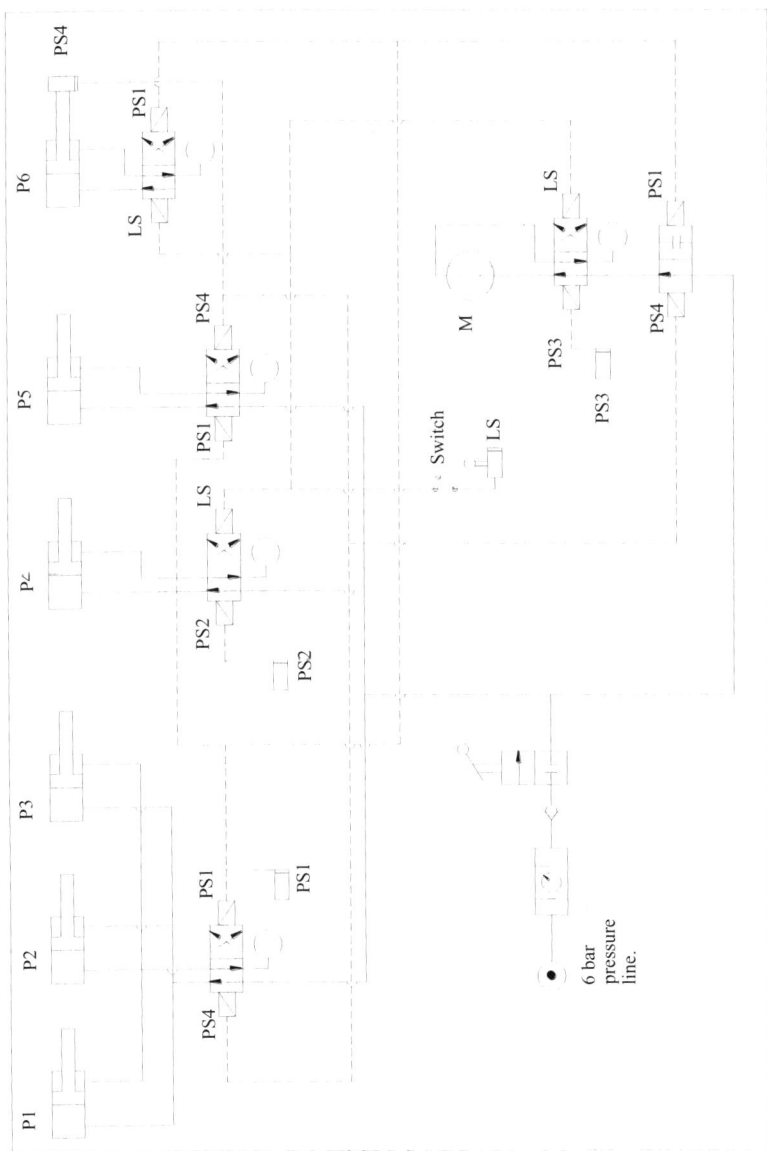

Figure 20: Pneumatic Ckt. diagram of Auto loader

7. TRIALS:

Several trials were taken manually to verify the feasibility of the auto loading system.

After the successful trials and approval from the management the project was offloaded to the KAIZEN work shop at BAL that deals with the continuous improvement projects for mounting the designed pneumatic circuit and further completion of the project.

8. TIME SAVING:

Time study using a stop watch was done for finding time required for the operator to unload parts from bin. This was done for 3W4S line and following were noted:

1. time required to reach the rack = 2.71sec
2. time required to load parts from bin to conveyor = 3.87sec
3. time required for placing empty bin on upper rack = 0.92 sec
4. returning to the destined work-station = 4.12sec

thus, total time required = 11.62sec

Since the operator has to spend 1.41sec in any case to reach the second machine, the total time required is = 10.21 sec

As per shift production is 580 jobs, operator has to load jobs for 72.5 times loading 8jobs each time. Thus per shift time saving is 740.225sec

9. ADVANTAGES:

1. The automation being manufactured in house was cheaper than any other flexible automation that was inevitable for the particular application.
2. Auto loading the parts on conveyor reduced the operators' fatigue.
3. The auto-loader project demanded few modifications in the existing methodology, hence could be easily implemented opening a window to future automation.
4. The design being dependent upon the bin size (which are standard) could be implemented with slight modifications parallely to almost all manufacturing lines.
5. The operators' saved times in loading jobs on conveyors could be utilized in more productive means.
6. The design incorporated KARAKURI which helped in reducing the cost of the project.

10. LIMITATIONS:

1. Number of reloading of jobs increased from 4 to 5 times per shift.
2. The automation involved vigorous and rapid movements of the auto-unloader unit which must be isolated keeping safety in view.
3. The reloading height on rack increased from 4ft to 6 ft requiring platform to be provided for reloading the rack.

11. CONCLUSION:

The low cost automation designed for the auto-loading of jobs to the conveyor line was found to be fulfilling all the previously stated objectives.

The autoloader designed was fool proof and could be parallely implemented in almost all cellular manufacturing lines, making small to medium sized components.

The project was offloaded to the "KAIZEN Workshop" at BAL for mounting pneumatic (circuits) components and completion of the project.

12. FUTURE SCOPE:

The low cost automation for auto loading of the components from bins to conveyor lines incorporated KARAKURI methods to ensure lower cost of the automation. The automation had many future scopes like:

1. Parallel implementation to all lines at BAL or any manufacturing firm having a cellular type layout.
2. Low cost automation for the entire manufacturing line using fixed automation arrangements for loading and unloading jobs from machines.
3. The automation would be of great use in robotized manufacturing lines where a separate operator needs to be employed for just emptying the bin and placing the jobs on conveyor from where robots pick up the parts. The auto loader unit would eliminate even the operator.

EFFECTIVE MATERIAL HANDLING SYSTEMS

1. INTRODUCTION

Material handling is the primary activity of every manufacturing organization. It has been estimated that at least 15 to 25% of the cost of the product is attributable to material handling activities.

Unlike many other operations, material handling adds to the cost of the product and not to its value. It is therefore important first to eliminate or at least minimise the need for material handling and second to minimize the cost of handling.

Material handling may be defined as the art and science of movement, handling and storage of material during different stages of manufacturing considered as material flow in to, through and away from the plant. It is in fact, the technique of getting the right goods safely to right place at right time and right cost.

Material handling in an organization takes place at various stages, such as the following:

- Unloading at goods inwards stores.
- Loading on to an internal transport.
- Movements to stores for the purpose of storage.
- Movement from stores to place of use(first work station)
- Movement to and from work stations.
- Movement to and from inspection bays.
- Movement to and from assembly benches.
- Movement to and from finished goods stores.
- Movement from and to dispatch department.
- Movement during packing.
- Loading of packed materials on to an external transport.

2. OBJECTIVE OF MATERIAL HANDLING:

A well planned material handling system should achieve the following objectives:

- Speed and economy in movement of materials (i.e. Minimization of processing time)
- Minimization of cost of material handling.
- Prevention of damages to materials.
- Safety in material handling (i.e. prevention of accidents)
- Minimization of fatigue and drudgery.
- Improvement in productivity.
- Higher plant efficiency.

- Greater utilization of material handling equipments.
- Better housekeeping.
- Efficient store keeping.
- Lower investment in work-in-process.

3. ENGINEERING AND ECONOMIC FACTORS:

Two important sets of factors to be considered in analyzing a material handling problem are:

- engineering factors
- economic factors

1. Engineering factors can be further sub grouped as under:

a) building and plant layout
b) manufacturing processes and equipments
c) nature of materials and product to be handled
d) material handling equipments

a) Building and plant layout

This factor considers features such as:

i. processes and departments to be tied
ii. width of aisles
iii. location of columns
iv. ceiling heights
v. number of floors
vi. load bearing strengths of the floors

b) Manufacturing process and equipment

This factor considers such features as:

i. production equipment
ii. method of production
iii. sequence of operations
iv. quantities of materials involved

c) Nature of materials and products to be handled

This factor considers features such as:

i. Nature of raw materials or parts handled (i.e. large or small, singly or together, heavy or light, symmetrical or non-symmetrical, rough or fragile etc.)
ii. quantities handled
iii. continuous or intermittent flow
iv. distances over which transported

d) Material handling equipment factors

This factor analyses features such as:

   i. Kind of equipment suitable for the job (e.g. trolleys, forklifts, trucks, conveyors, overhead cranes etc.)

   ii. capacity of equipment

   iii. hours of service per day

   iv. space required for operation

   v. power requirements

   vi. ease of operations

   vii. speed of operation

   viii. auxiliary equipment required

   ix. adaptability with the other equipments in use or contemplated

2. Economic factors

Economic factors analyses features such as

   i. initial cost of equipment

   ii. cost of installation including rearrangement or alterations to the existing equipment

   iii. cost of maintenance and repairs

   iv. cost of power

   v. cost of labour required to operate the equipment

   vi. taxes and insurance

   vii. interest on investment

   viii. depreciation

   ix. license fees (e.g. trucks)

   x. supervision costs

   xi. salvage value

   xii. saving due to reduction in number of men released for other work

   xiii. saving due to expenses on equipment displaced

   xiv. savings due to reduction in rework and rejection on account of improvement in handling

   xv. savings due to increase in production as a direct consequence of changes in materials handling system

4. PRINCIPLES OF MATERIALS HANDELING:

1) OPERATING PRINCIPLES

   a) "Unit load handling" principle

   "Material handling cost is inversely proportional to the size of load (unit load)."

Unit load implies packaging of several pieces or arranging of several pieces on a skid, pallet or platform for movement as single unit. Basic principles of unit load handling are

- Material should be handled in bulk/ lots over distances. It is better to wait unless otherwise there is reason unless there is a trolley or barrow load before moving them instead of having a laborers carry each one separately.
- Fragile or breakable materials should be arranged in trays or in layers separated by strapping

b) "GRAVITY" principle

"Gravity is the most economical motive force. Materials, whenever possible, should be moved using this nature's greatest resource."

Gravity principle suggest that

- Material whenever possible should be made to roll or slide down the chutes to the next work station instead of pushing it or carrying it.

c) "Flow of materials" principle

"Materials handling efficiency is the greatest when it approaches a steady flow of materials, in a straight path as possible, with minimum of interruptions and minimum of back tracking."

That is:

- Material should be made to move over as straight path as possible it cut down distance.
- Re-handling and cress-cross movements should be avoided.

2) EQUIPMENT PRINCIPLES

a) "Mechanization" principle

"Mechanization of materials handling (the use of mechanized equipment instead of man power) generally increases efficiency and economy in handling."

The use of mechanized equipments instead of manpower reduces handling cost, increases speed of handling, cuts down accidents, improves safety, reduces fatigue of the workers and improves productivity since work done by power is cheaper and large volume of work can be handled. One should look into the following:-

- Heavy materials should be lifted mechanically and moved mechanically. Manually handling of heavy materials can cause sprains and certain diseases like hernia etc.
- Manual handling makes the firm to commit itself to a continuous expenditure by way of labour wages which through mechanization goes down drastically. However, materials handling equipments cost lot of money and therefore pros and cons must be weighed carefully before arriving at the decision

- Mechanization of all operations may not be necessary. Only a part of material handling operations may be mechanized. Areas where this can be done economically are conveyors in process industries or those industries where the same product is produced continuously.

b) "Terminal time" principle

"Reduction in terminal time of handling equipment increases efficiency and economy of the equipment." This principle may be amplified to state that:

- Waiting time of the equipment at the pick up and put down points should be reduced to its minimum by cutting down loading and unloading time.
- Palletizing and unit load concept should be used to reduce terminal time.

c) "Dead weight" principle

"Economy of the equipment is directly proportional to the ratio of load handled to the dead weight of trays, trolleys and pallets should be the least possible."

To improve efficiency of the material handling equipment:

- Dead weight of the equipment should be reduced to its minimum.
- Weight of trays, trolleys and pallets should be the least possible.

d) "Standardization " principle

"Standardization of material handling equipment increases efficiency and gives economy in operation of the equipment."

Standardization benefits the firm in more than one way It:

- Permits interchangeability of equipment between departments.
- Reduces investment in spares inventory as few parts require to be stocked.
- Reduces maintenance and repairs cost.

e) "Maintenance" principle

"Systematic maintenance, planned repairs and replacements, inspection and adjustments, lubrication and cleaning activities etc. increases efficiency and productivity of materials handling equipment."

Material handling equipment cost lot of money and many a times in short supply. The equipments, therefore, must be made available for maximum possible time to render services to the production function. Production suffers if the equipments fail abruptly. To get trouble free service:

- Repairs and replacements must be anticipated.
- Planned preventive maintenance program should be implemented for all handling equipments.

f) "Speed" principle

"Economy in material handling increases with efficiency and speed of material handling."

For this, the following points may be considered:

- Pallets should be made square in shape so that forks can enter them from any side. This reduces pick up time.

- Two way traffic routes may be followed to eliminate "empty run".

- Gangways should be kept clear. It is no use investing in expensive handling equipment if it is going to be held up by obstructions.

- Number of boxes, pallets, or containers available at the work place should be sufficient to eliminate waiting by material handling equipment.

g) "Versatility" principle

"Economy in material handling is obtained by use of equipment that is capable of variety of applications."

Special equipments with limited range of use will only be economical if their total time is justified and further changes in plan are not foreseen. In all other cases, equipment with variety of uses should be preferred. The difficulty will not be experienced by larger plants as they usually have variety of equipment but will be felt by medium and small plants.

3) COSTING PRINCIPLES

a) " Equipment selection" principle

"Selection of the most flexible equipment after thorough study of the items and materials to be moved increases efficiency of material handling."

Selection of material handling equipment is a vital decision. Many a factors require to be considered in the selection of the equipment.

b) "Replacement" Principle

"Material handling cost is the lowest if the equipment is used only for its economic retentive period and is replaced by an alternate based on engineering economic principles."

Material handling equipment like production machines too has a certain economic life. Retaining equipment beyond its economic period increases repairs cost and causes production holds up due to eventual breakdowns.

c) "Handling cost appraisal" principle

"Periodic analysis of materials handling costs highlights areas of improvements."

4) GENERAL PRINCIPLES

a) "Safety" principle

"Materials handling efficiency increases as working conditions are made safer and safer." That is-

- Hot materials should be handled and moved mechanically.
- Gangways should be kept lighted and unobstructed to avoid damages and injuries.
- Operators must be well trained to handle equipment.
- Aisles should be wide and unconguested.
- Workmen should be provided protective clothing (wherever necessary.)
- Safety regulations, statutory or otherwise, should be strictly adhered to.

b) "TRAINING" PRINCIPLES

"Training of workmen in good material handling techniques and systems and educating them towards importance of their work usually helps to develop right attitude to material handling."

This principle suggests that:

- Each employee should be given basic training in material handling techniques.
- Each employee should be told the effect of material handling on product quality and product cost.
- Work culture should be developed in such a way that workman after performing the operation on a job keeps it in a position that is ready to be moved on.

c) "IDENTIFICATION" PRINCIPLES

"Materials must be kept identified by labeling on pallets and boxes."

This is important failing which unlabelled boxes may require to be opened to look for what they contain and also when certain required materials are not traceable.

d) "Location" Principle

"All handling equipments should be placed at the right place and at the right time to avoid hunting and delays in material handling cost of the materials."

- To hunt for a materials handling equipment, be it a trolley, helper, or truck, or any other, wastes time and retards flow of materials. There should be a place for everything.

e) "MATERIAL TREATMENT" PRINCIPLE

"All types of materials should be treated as important since material handling costs are not related to the cost of materials."

This principle underlines the fact that

- Scrap and end pieces are as important for material handling as production materials.

45

- Low value bulky materials cost much more than high value compact materials. Materials handling costs are not related to the cost of material but are dependent upon their bulk and physical and chemical characteristics.

5. EXTSTING WAYS OF MATERIAL TRANSFER BETWEEN MACHINES IN MANUFACTURING LINES AT BAL

In BAL various types of material handling techniques are used suiting to the type of process done on the jobs.

Material transfer in manufacturing lines involves transfer of parts / jobs from one station to other for performing various manufacturing operations on the job in a predetermined sequence.

This is done by following methods:

1. Operator carrying jobs manually from station to station.
2. Use of powered conveyors for material transfer.
3. Using gravity chutes or roller conveyors for material transfer.

1. Operator carrying jobs manually from machine to machine:

As shown in diagram, operator of machine 1 after machining the job picks the job from the fixture of machine 1 and keeps it on job collector. Then the operator on

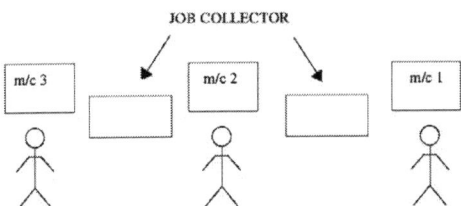

Figure 21: Job Collector

machine 2 collects the job from the job collector and loads it on machine 2 for subsequent operation. The chain continues throughout the line. Such type of material transfer is observed at the end of U-lines, gear manufacturing shops, heat treatment shops.

Job collector may be of the following types:

- Bins stalked in between two machines/stations.
- Trays between two machines/stations.
- Special trolleys placed between two machines/stations.

46

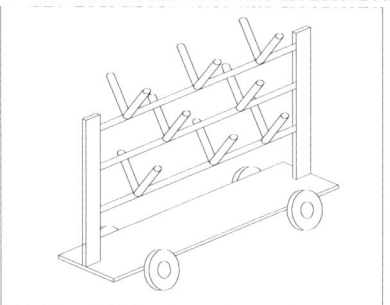

Figure 22: Special troly for Mateial Handling

ADVANTAGES:-

1. Material transfer manually is less costly for small sized jobs as cheap labour is available.

2. Absence of any power consuming devices for material transfer reduces electricity costs.

3. Machines can be used to the fullest of their capacity, to produce buffer stock of materials.

4. Low investment in material transfer equipments.

DISADVANTAGES:-

1. Heavy and large jobs cannot be handled with ease.

2. Speed of material is low and accident prone.

3. Non productive time increases as operator's valuable time unnecessarily wasted in material transfer.

4. Manual material transfer causes fatigue of the operator.

2. Use of Powered Conveyors for material transfer:

Powered Conveyor is a fixed type of material transfer system driven by some prime mover, used for moving material either continuously or intermittently between two fixed points (in either vertical or horizontal direction.)

Conveyors are of various types, namely belt conveyors, roller conveyors, bucket conveyors, screw conveyors, monorail conveyors, troughed conveyors, vibrating conveyors, oscillating conveyors, etc.

CONVEYOR

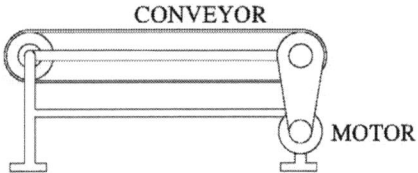

MOTOR

Figure 23: Powered conveyors

- **Belt Conveyors:** Belt conveyors are flexible and can be designed to handle almost everything. Belt conveyors are used to carry objects over short as well as long distances and to convey goods to the upper floors. Belt conveyors generally move slowly (25-30 meter per minute) but some move at high speeds (200 meter per minute or more). They are used for picking, sorting delivering materials move at slow speed (3-15 meter per minute) and those used for cleaning , painting, baking, drying and cooling move still slowly (perhaps 0.5-2 meter per minute).

- **Screw Conveyors:** They make use of a propeller like helical fin or screw to push materials in one direction. Screw conveyors are used to handle bulk materials and only those which are not damaged by crushing.

- **Monorail conveyors:** They use hooks, racks, baskets, or dollies to transport a variety of objects. The overhead monorail conveyors use a close –circuit-track to ride parts in space and bring past the point of use at regular intervals.

- **Vibrating conveyors:** They employ metal troughs, jerk quickly towards the fixed end and return back to repeat at high frequency, almost 1000 times a minute. They are used to transport materials over the short horizontal distances (normally less than 30 meter).

- **Oscillating Conveyors:** Like vibrating conveyors, they use metal troughs but jerk less frequency around, 200-250 times per minute, to transport materials steadily to take-off end Powered conveyors are widely used in straight line flow type manufacturing lines and assembly lines. In BAL use of powered conveyors can be witnessed in Press shop, aluminium components machining shop, steel machine shop and engine & vehicle assembly lines.

ADVANTAGES:

- Conveyors create relatively fixed route. Materials move continuously as required in continuous manufacturing.
- They occupy space continuously unless they are of portable type.
- They reduce handling of jobs and therefore are fast and safer in operation.
- Suitable to manufacturing plants where the rate of movement of material and unit load are fixed.
- They reduce operator fatigue and non productive time.
- Heavy and awkward sized jobs can be easily transferred.

DISADVANTAGES:

- Initial investment on conveyor equipment is high.
- Consume electricity hence increasing electricity expenses.
- Application restricted to fixed routes.
- Regular maintenance is required.

3. Gravity chutes and roller conveyors for material transfer

Sliders or roller conveyors are provided between two machines in this type of material transfer.

- **Roller Conveyors:** They are made of metal frame work bearing horizontal rollers placed at intervals. They use rollers gravity concept and are placed between two destinations so that goods can be pushed along the top of the rollers.

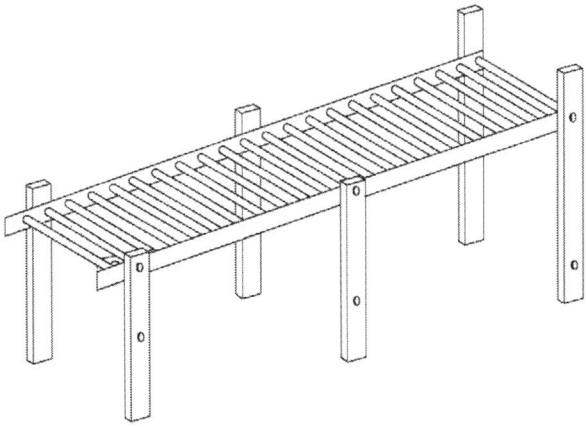

Figure 24: Gravity Roller Conveyor

- **Gravity chutes:** They are generally made up of steel sheets with walls to guide the jobs. The chutes are provided with slope to utilize the gravity to transfer jobs between two destinations.

In BAL gravity chutes are used in Press shop, rollers are used in aluminium machine shops for transferring jobs between subsequent machining stations. Rollers racks are used for transferring bins having jobs in them.

ADVANTAGES:

- Low initial investment on material transfer equipments.
- Absence of any power consuming devices for material transfer reduces electricity costs.
- No maintenance is required for this type of material transfer.
- They reduce operator fatigue and non productive time.

DISADVANTAGES:

- Using chutes can increase job rejection as jobs rub on slider surfaces during transfer.
- Pushing the job to the next station causes operator fatigue.
- This type of material transfer is not suitable for heavy jobs and between distant stations.

6. KARAKURI; Concept and need for improved material handling

Karakuri is a Japanese terminology meaning "Puzzle". It is one of the oldest techniques used for manufacturing toys or the things for amusement and entertainment. Karakuri techniques utilize gravity and forms of spiral springs as energy sources with the help of mechanical linkages and their interlocks. This method requires indeed a great creativity and imagination.

Moving toys operating on simple / complex mechanical linkages from the past to the recent times toys like the car reversing when it bumps / dashes against surface are some very common examples of the Karakuri.

Recently the industries have started thinking and developing upon the Karakuri techniques for effective material handling. *Karakuri Kaizen* employs simple techniques to reduce human effort and simplify processes.

Considering the advantages and disadvantages of the existing material transferring practices in BAL and understanding the principles of Karakuri, a scope of improving the material handling systems was seen. The effective material handling system would:

- Reduce electricity consumption
- Reduce operator's fatigue and drudgery
- Prevent of damage to materials.
- Higher plant efficiency
- Minimization of cost of material handling.
- Use gravity, i.e. counterweights

DESIGN, FABRICATION AND IMPLEMENTATION OF POWERLESS MATERIAL HANDLING EQUIPMENTS

(A) POWERLESS CONVEYORS

1 PROJECT OBJECTIVE

In manufacturing lines material transfer commonly took place either by powered conveyors, roller conveyors, gravity chutes or manually. As stated above these types of material transfer systems had the following drawbacks:

i) Powered conveyors involve high initial investment, running and maintenance cost.

ii) Roller conveyors requires pushing the job on conveyor resulting in operators fatigue, moreover they have considerable initial investment.

iii) Gravity chutes may result in job rejection due to rubbing of jobs on chute.

iv) Manual handling increases non production time and causes operator fatigue.

To overcome these drawbacks powerless conveyors were needed to be designed.

Thus POWERLESS CONVEYORS aim towards replacing the electrically powered or chutes and roller conveyors used in manufacturing lines.

2. CONCEPT AND WORKING

The POWERLESS CONVEYOR uses gravity as the energy source.

The figure demonstrates the concept of the powerless conveyor.

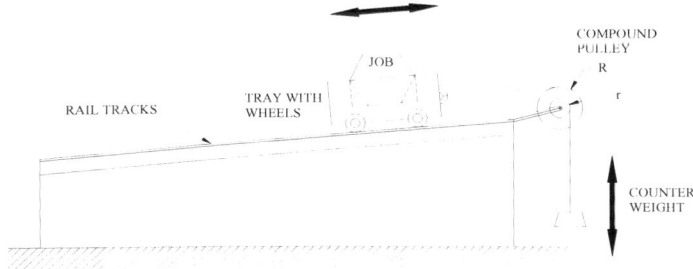

Figure 25: Concept of Powerless Conveyor

As shown in the figure the powerless conveyor consists of a tray/ pan with wheels which roll on the rails that are provided on the frame. The tray is tied to a string which rolls over a pulley that is compounded with a smaller pulley. A string rolls over the smaller pulley and is tied to a counterweight.

The counterweight is such that it balances the tray at the top most position on the track. When a job is kept on the tray at the top most position it moves down thus decreasing its potential energy against the increasing potential energy of the counterweight. The system is designed such that the equilibrium between potential energy of the tray with job and counterweight is reached when the tray reaches the bottommost position of the tray. The radius of pulleys is selected such that the tray can travel the required distance for the available height restriction to the counterweight.

MATHEMATICAL ANALYSIS

Let,

> m_1 be the mass of empty pan in kg
>
> P_w be the weight of parts in N
>
> m_d be the mass of counterweight needed to be attached in kg
>
> θ be angle of inclination of the conveyor track.
>
> μ be the coefficient of friction between the track and roller bearings of pan
>
> R be radius of the larger pulley
>
> r be the radius of the small pulley
>
> L be the distance traveled along the path of pan
>
> h be the distance traveled in vertical direction

Also,

$$h = (2 \, \pi \, r) \, n \qquad \text{... (1) a}$$

And $\quad L = (2 \, \pi \, R) \, n \qquad \text{... (1) b}$

> Where, n is the no. of revolutions made by the pulley

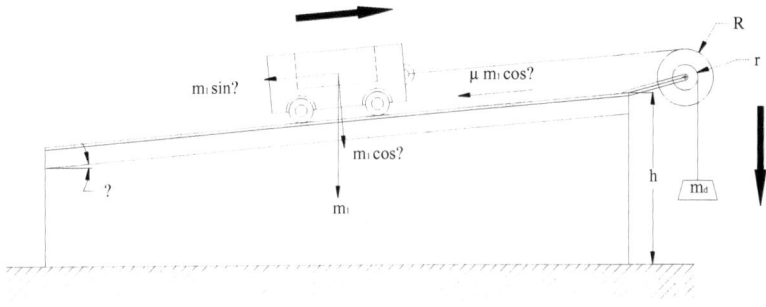

Figure 26: FBD of Tray while moving up

(i) Consider the FBD for empty bin during upward motion due to counterweight,

$$(m_1 \sin\theta + \mu \, m_1 \cos\theta) \, g \, (2 \, \pi \, n \, R) = m_d \, g \, (2 \, \pi \, n \, r)$$

i.e. $m_1 (\sin\theta + \mu \cos\theta) \, R = m_d \, r \qquad \text{... (2)}$

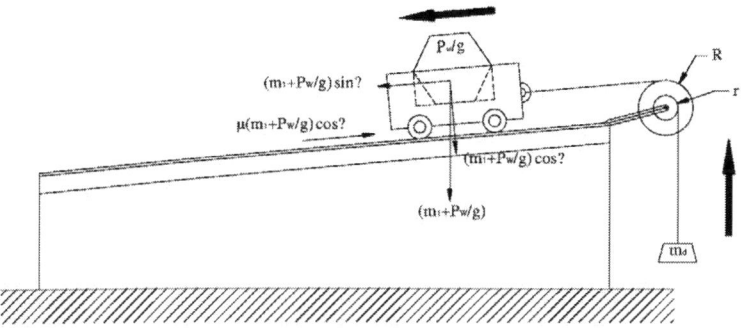

Figure 27: FBD of tray while moving down

(ii) consider the FBD for bin during downward with the part(s) against the counterweights,

$$[(m_1 + P_w/g) \sin\theta - \mu (m_1 + P_w/g) \cos\theta] g (2 \text{ л n R}) = m_d g (2 \text{ л n r})$$

i.e. $(m_1 + P_w/g) (\sin\theta - \mu \cos\theta) R = m_d r$ ... (3)

Equating (2) & (3), we get,

$$m_1 (\sin\theta + \mu \cos\theta) R = (m_1 + P_w/g) (\sin\theta - \mu \cos\theta) R$$

i.e. $(m1 + P_w/g) / m_1 = (\sin\theta + \mu \cos\theta) / (\sin\theta - \mu \cos\theta)$

i.e. $(m1 + P_w/g) / m_1 = (1 + \mu \cot\theta) / (1 - \mu \cot\theta)$

i.e. $(P_w/g) / m_1 = 2 \mu \cot\theta / (1 - \mu \cot\theta)$

Hence,

$$\cot\theta = 1/\mu [(P_w/g) / (2 m_1 + P_w/g)]$$ ... (4)

Using equations,

- 1 (a) &1 (b), the radii of pulleys can be found out if length of track and height of conveyor are known.

- 4, the angle of inclination of the conveyor can be found, if the mass of empty pan and mass of parts are known.

3.   DESIGN

General steps for design of powerless conveyor:-

From the above analysis, following steps are derived to design the counterweight operated conveyor for any application.

1. Decide the height (h) and length (L) of the conveyor suiting the application.
2. Select the bigger pulley from the available, of suitable radius (R) and calculate the number of revolutions (n) using equation 1 (a).
3. Calculate the radius of smaller pulley (r) using equation 1 (b).
4. Measure the weight of the empty pan (m_1) and parts (P_w/g) to be conveyed. Using the values calculate the angle of inclination (θ) of conveyor using equation 4.
5.   Calculate counterweight (m_d) from above values using equation 2.

DESIGN OF COUNTERWEIGHT OPERATED CONVEYOR

The counterweight operated was designed for K-70 Cylinder head line.

And the average distance between two consecutive machines was measured to be 2m (2000 mm)

i.e. L = 2000 mm

The standard height of conveyor was measured to be 934mm and the same height was adapted,

i.e. h = 934 mm

as per the space available a suitable value of R was selected

R = 50mm

From equation 1(a) & 1(b)

$$n\, 2\, \pi\, R = 2000$$

i.e. $n\, 2\, \pi\, 50 = 2000$

i.e. $n = 6.366$

$$\approx 7$$

also, $n\, 2\, \pi\, r = 934$

i.e. $7 \times 2\, \pi\, r = 934$

i.e. $r = 21.2$

= 22mm.

Mass of K-70 cylinder block was measured to be 0.8 kg, so designing for two parts at a time P_w/g was taken to be 1.6 kg

And the mass of empty pan (m_1) was measured to be 0.5 kg

Hence, angle of inclination was calculated using equation 4,

$$\cot\theta = 1.6 / (2 \times 0.5 + 1.6)\mu$$

The value of coefficient of friction (μ) for hardened ground steel and mild steel is

$\mu = 0.06$

hence, $\cot\theta = 0.0975$

i.e. $\theta \approx 5^{\circ}$

The counterweight m_d was calculated using equation 2,

$$m_1 (\sin\theta + \mu \cos\theta) R = m_d r$$

i.e. $0.5 (\sin 5 + 0.06 \cos 5) 50 = m_d \times 22$

i.e. $m_d = 0.778$ kg

Thus the designed parameters are:

- height of conveyor, h = 93.4 cm
- length of conveyor, L = 3000cm
- Radius of pulleys , R = 5 cm

  r = 2.2 cm
- Angle of inclination of conveyor $\theta = 5^{\circ}$
- Mass of counterweight $m_d = 0.778$ kg

4. FABRICATION AND IMPLEMENTATION

The powerless conveyor was fabricated in the central maintenance of BAL.

The components/parts used in above designed powerless conveyor were fabricated as follows:

- TRAY :

  Material used – 3mm steel sheet

  Dimensions of the tray 26 X 18 X 5 cm

QTY - 1

The steel sheet of 36 X 28 was cut on shearing m/c then squares of 5 X 5 cm was cut out from the corners on nibbling m/c. the tray was then made out of the cut shape by bending it on bending m/c. the corners were spot welded.

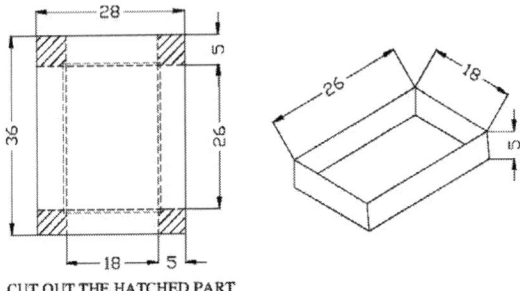

CUT OUT THE HATCHED PART

Figure 28: Tray Dimensions

- COMPOUND PULLEY :

    Material – mild steel

    Dimensions - As Shown Below

    QTY – 1

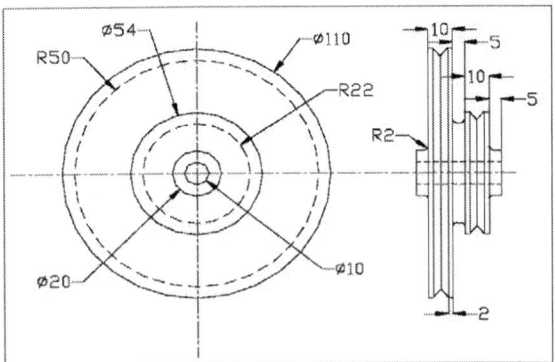

Figure 29: Compound Pulley

- WHEELS :

    Radial bearings – o.d = Ø 30 mm

    i.d = Ø12 mm

    QTY – 4

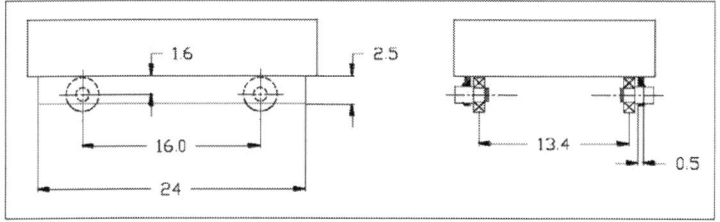

Figure 30: Tray with wheels

- PINS :

 (I)   Material – mild steel

Dimension – STP Ø12 / Ø15 x 30 mm

QTY – 4

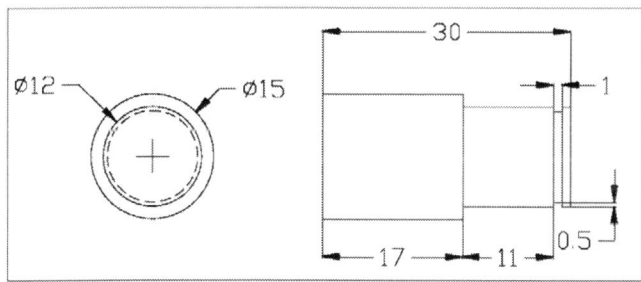

Figure 31: Pin

 (II)   Material – mild steel

Dimension - Ø10 X 100

QTY – 1

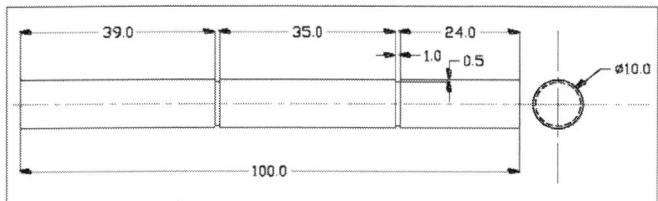

Figure 32: Axle

- FRAME :

Material – 50mm square channels

3mm steel sheets

Mild steel rods of dia 10mm

Dimension – as shown below

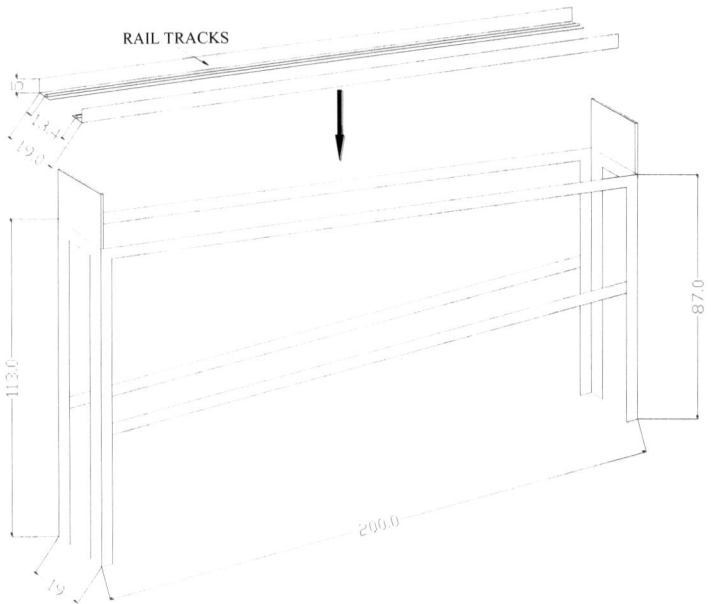

Figure 33: Frame: Powerless Conveyor

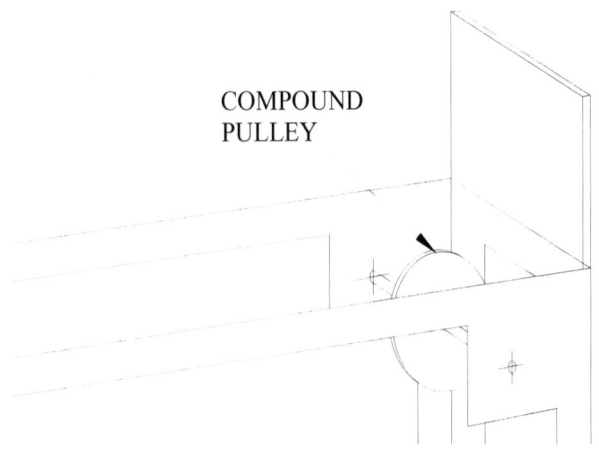

COMPOUND
PULLEY

- COUTERWEIGHT :

MS bar of 80 mm diameter and weighing 780 gm as per designed dimension was used as counter weight. A hook was weld on the counterweight for tying the string.

After fabrication of all the components required, they were assembled and the powerless conveyor was installed in the K 70 cylinder head line. Several trials were conducted and the equipment was found to perform consistently and satisfactorily.

5. ENERGY SAVINGS FOR K70 CYLINDER HEAD LINE:

The power conveyor used on K 70 cylinder head line consists of three electrics motor each of 500 watts. Hence power consumed in a shift of 420 minutes is give as follows:

Power consumed = total power required x time of operation

$$=1.5 \text{ kw x } 420/60$$
$$= 9 \text{ kw-hrs}$$
$$= 378 \text{ joules}$$

Thus replacing the existing powered conveyor with powerless conveyor will save 9 kw-hrs per shift.

Figure 34: Photographs of Powerless Conveyors

6. ADVANTAGES:

- Material transfer with absolutely no power consumption.
- Cost effective, as it involves low fabrication, running and maintenance cost.
- Automatic material transfer after the operator just places the job on the tray.
- Part handling is safe, as use of suitably sized trey does not damage the part.
- Following simple steps appropriate systems can be designed to suit the application.
- The system can also be used for material transfer in curved paths.
- Free from chip accumulation problems.
- The simple and robust construction ensures long life of the equipment.
- Suitable for complete automation of manufacturing lines.

7. LIMITATIONS:

- The conveyors are suitable for short or medium distances as longer length of conveyor causes greater reduction of operating height due to slope.
- Can not be used in manufacturing lines having machines with different machining time.
- New system needed to be designed every time for varying conveying lengths and for new parts of different weights.

(B) Revolving material transfer system

1. PROJECT OBJECTIVE

It is always advisable to minimize or if possible eliminate operator's involvement in transferring materials. There was a need to develop a simple, compact and robust material handling equipment that would suit almost any short distance material transfer task. Such a system would eliminate the need of powered conveyors, chutes or roller conveyors that need operator attention and picking of jobs by hand to transfer it between two stations.

The system would be of great use in manufacturing lines having U type or Zig Zag type layouts where human handling of jobs is presently unavoidable, as use of powered conveyors or gravity conveyors cause interruption in line operations.

Thus the objective of the project was to develop Material transfer equipment that:

- Would eliminate the operators handling of jobs for transferring them between stations required in U type, circular type, or Zigzag manufacturing lines.
- Would be simple, robust and compact, and would suit almost any short distance material transfer task
- Would work on the principles of *karakuri*, and would consume no electrical power.
- Would result in increase of effective productive time
- Would require negligible maintenance attention.

2. CONCEPT AND WORKING

Taking all the objectives into consideration a REVOLVING MATERIAL TRANSFER SYSTEM was conceptualized.

"THE REVOLVING MATERIAL TRANSFER SYSTEM" works on the principle of weights and counterweights.

The figure shown below demonstrates the concept of Revolving Material Transfer System.

Figure 35: Concept of Revolving Tray

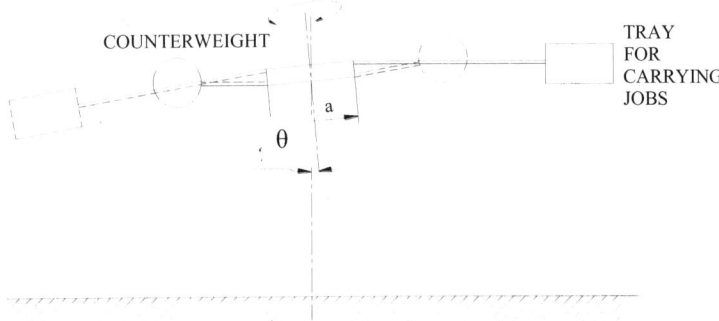

As shown in the above figure, the tray that carries the job is connected to a rotating assembly by means of a telescopic arm. A counter-weight is attached diametrically opposite to the tray arm. Both the arms, i.e., the tray arm and counter weight arm are free to revolve about the central axis. The entire revolving assembly is pivoted with its axis of revolution having a small inclination with the vertical.

The counter-weight is selected such that it balances the weight of the empty tray at the topmost position (assuming negligible weight of arms). When the tray is empty system has the tray arm at the top most position. When the part is kept on the tray the combined weight of the tray and part exceeds the counterweight and thus tries to attain a lower potential. This causes the tray to revolve by 180° about the axis of rotation, which results in transfer of the job between two stations.

Once the part reaches the destined station and is lifted from the tray, the counterweight side becomes heavier than the tray side. So now the counterweight which had reached its maximum attainable potential tries to return back to its lowest potential, causing the assembly to revolve again to acquire the initial positions.

Hence the revolving tray moves from first station to second when a job is placed on it and returns back automatically to the previous station once the job is removed from the tray. The system may work for any angle between 0° to 180°.

MATHEMATICAL ANALYSIS:

Figure 36: FBD of Empty Revolving tray

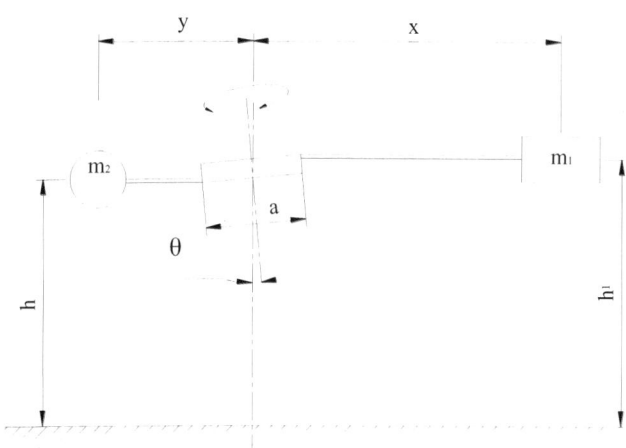

Let,

m₁ be the mass of empty pan in kg

P_w be the weight of parts in N

m₂ be the mass of counterweight needed to be attached in kg

θ be angle of inclination of the conveyor track.

x be the length of tray arm from the axis of rotation in mm

y be the length of the counterweight arm from the axis of rotation in mm

a be the diameter of the bearing housing in mm

Consider,

Position 1 i.e., when the tray is empty before rotation

h_1 be the height of the pan from datum in mm

h be the height of the counterweight from datum in mm

Position 2 i.e., after rotation when the tray carries job

h_1' be the height of the pan from datum in mm

h' be the height of the counterweight from datum in mm

Taking moments about the axis of rotation we can write,

$(m_1 \ h_1 \ g) \ x = (m_2 \ h \ g) \ y$

From the fig, using geometry

$$h_1 = h + a \sin\theta$$

Substituting h_1 in the first equation, we get,

$$m_1 (h + a \sin\theta) g x = m_2 y g h \qquad \dots 1$$

therefore, $(m_2 y) / (m_1 x) = (h + a \sin\theta) / h$

i.e. $(m_2 y) / (m_1 x) = [1 + (a / h) \sin\theta]$

i.e. $\sin\theta = [(m_2 y - m_1 x) / m_1 x - 1] h / a$

Thus, $\qquad \theta = \sin^{-1} (h / a [(m_2 y - m_1 x) / m_1 x]) \quad \dots 2$

When the job is added on the tray the necessary condition for the system to work is,

$(m_1 + Pw/g) (h + a \sin\theta) g x > m_2 h y g + f$

Where, f is the frictional resistance experienced in the bearing during rotation.

Now to ensure complete 180° revolution of the tray about the axis, the following condition is necessary to be fulfilled.

$(m_1 + Pw/g) (h + a \sin\theta - x \sin2\theta) x \geq m_2 (h + y \sin2\theta) y + f$

Neglecting f, as use of bearings offer negligible resistance in motion, also since a ≪ h;

$$a \sin2\theta \approx 0$$

hence,

$(m_1 + Pw/g) (h - x \sin2\theta) x \geq m_2 (h + y \sin2\theta) y$

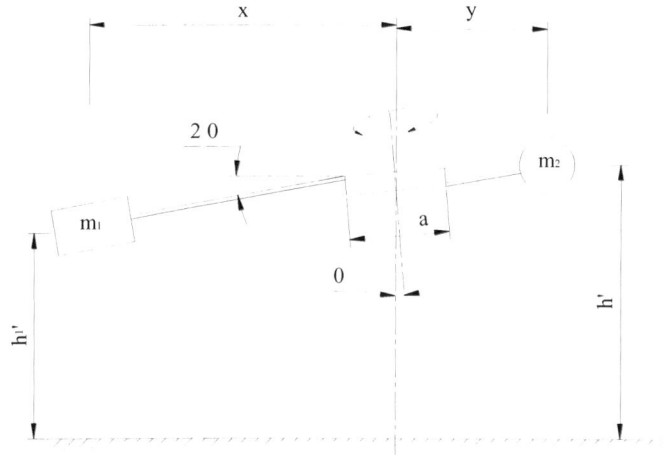

Figure 37: Revolving Tray with Part

i.e. $(m_1 h x) + Pw/g x h - m_1 x^2 \sin2\theta - Pw/g x^2 \sin2\theta \geq m_2 h y + m_2 y^2 \sin2\theta$

i.e. $Pw/g \geq (m_1 x^2 + m_2 y^2) \sin2\theta / [x (h - x\sin2\theta)]$ ...3

As shown in the above condition, $(m_1 + Pw/g)$ will try to come to the lower potential by converting its potential energy into kinetic energy by traveling to a lower height $(h - x \sin2\theta)$, $[(h - x\sin2\theta)$ being the height corresponding to the lowest potential available for $(m_1 + Pw/g)$ in the system]. Consequently it will take m_2 to the height $(h + y\sin2\theta)$ thus converting the kinetic energy into potential energy $[(h + y\sin2\theta)$ being the highest potential for m_2 in the system]

Thus for the above condition, $(m_1 + Pw/g)$ & m_2 will revolve around the axis through $180°$.
If,

$Pw/g < (m_1 x^2 + m_2 y^2) \sin2\theta / [x (h - x\sin2\theta)]$

The system will not revolve to complete an angle of $180°$ but attain equilibrium at some intermediate angle say Φ.

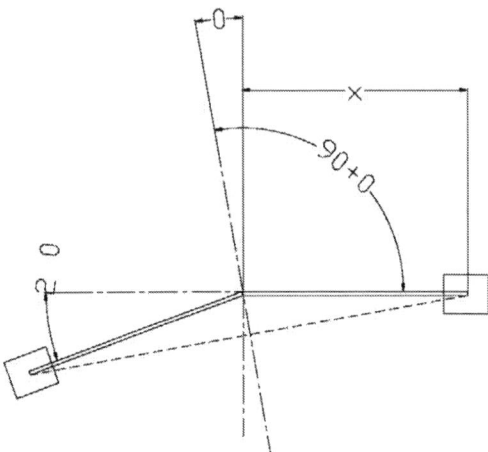

Figure 38: Cone of Action

Φ (Rotation through radians)	Distances	Variable	
• 0 to π/2	0 to x sinθ cosθ	x	
	0 to y sinθ cosθ		y
• π/2 to π	x to x sin2θ	x	
	y to y sin2θ	y	
• 0 to π	0 to xsin2θ	x	
	0 to ysin2θ		y

Thus the equilibrium of the system for above condition will be at angle Φ where height of tray h'$_1$ and height of counter weight h' will be given by

h'$_1$ = h − (Φ/π) x sin2θ

h' = h + (Φ/π) ysin2θ

Taking moment about central axis for the equilibrium position, we get

$(m_1 + Pw/g) [h − (\Phi/\pi) \times x \sin2\theta] x = m_2 [h + (\Phi/\pi) \times y \sin2\theta] y$

i.e. $m_1 x h − m_1 x^2 (\Phi/\pi) \sin2\theta + Pw/g h x − Pw/g (\Phi/\pi) x^2 \sin2\theta$

$= m_2 h y + m_2 y^2 (\Phi/\pi) \sin2\theta$

i.e. $Pw/g h x = (m_1 + Pw/g) (\Phi/\pi) x^2 \sin2\theta + m_2 (\Phi/\pi) y^2 \sin2\theta$

i.e. $\quad Pw/g \, h \, x = (\Phi/\pi) \sin2\theta \, [(m_1 + Pw/g) \, x^2 + m_2 \, y^2]$

Hence $\quad \Phi = \pi \, Pw/g \, h \, x \, / \, \{\sin2\theta \, [(m_1 + Pw/g) \, x^2 + m_2 \, y^2]\} \qquad \dots 4$

Special cases:

Deriving mass of part/job required for system to revolve through Φ

For $\quad \Phi = \pi^c \qquad Pw/g = (m_1 \, x^2 + m_2 \, y^2) \sin2\theta \, /[x \, (h - x\sin2\theta)]$

$\Phi = \pi/2^c \qquad Pw/g = (m_1 \, x^2 + m_2 \, y^2) \sin2\theta \, /[x \, (2h - x\sin2\theta)]$

$\Phi = (\pi/S)^c \qquad Pw/g = (m_1 \, x^2 + m_2 \, y^2) \sin2\theta/[x \, (S \, h - x\sin2\theta)]$

$\qquad\qquad\qquad = k1/(S \times k3 - k2)$

Where, k1, k2, k3 and S are constants.

RESULTS:

i. The part will revolve around the tilted axis through 180° when

$\theta = \sin^{-1} \{(h/a) \, [(m_2 \, y - m_1 \, x)/ \, m_1 \, x]\}$

and $Pw/g \geq (m_1 \, x^2 + m_2 \, y^2) \sin2\theta \, / \, [x \, (h - x\sin2\theta \,)]$

ii. The part will revolve through Φ^c around the tilted axis when

$\theta = \sin{-1}\{(h/a) \, [(m_2 \, y - m_1 \, x)/ \, m_1 \, x]\}$

and $Pw/g < (m_1 \, x^2 + m_2 \, y^2) \sin2\theta \, /(x \, [h - x \, \sin2\theta])$

where Φ^c is given by

$\Phi = \pi \, Pw/g \, h \, x \, / \, \{\sin2\theta \, [(m_1 + Pw/g) \, x^2 + m_2 \, y^2]\}$

iii. Also if $Pw/g = (m_1 \, x^2 + m_2 \, y^2 \, / \, ([S \, h - x\sin2\theta \quad]x)$ the angle through which the part will revolve Φ is given by

$\Phi = (\pi/S)^c$ where S is any integer

3. DESIGN

General steps for designing the revolving material transfer system

1. Decide the working height h of the REVOLVING MATERIAL TRANSFER SYSTEM, and the length of the tray arm x, suitable to the application.
2. Determine the diameter of the bearing housing a and fix a length of the counterweight arm y.
3. Measure the weight of the part(s) Pw/g to be transferred. And determine the weight of the tray & tray arm m_1.
4. Using equation 1 determine the counterweight m_2.
5. Using equation 2 calculate the angle of inclination θ.

DESIGN FOR CYLINDER HEAD 3W4S

The 3W4S cylinder head manufacturing line has a U type layout as shown in the figure. For the material transfer at the end of the U line the operator had to carry the washed components from washing machine to the leak testing station on the opposite of the washing machine. This w
as a suitable location for implementing the Revolving material transfer system.

After taking measurements of the location the tray arm length was decided to be

X =1.2 m =1200 mm

According to the principles of ergonomics the Height of the system was decided to be, h
= 934 mm

The diameter of bearing housing was taken to be a = 150 mm

The counterweight arm was taken to be y = 300 mm

The weight of tray and tray arm was assumed to be = 1 kg

To compensate for the shift of cg of the tray side due to significant weight of the tray arm the value of 'x' was taken to be x = 0.9 m = 900 mm

Hence using equation 1, counterweight was calculated as,

$m_1 (h + a \sin\theta) g x = m_2 y g h$

Neglecting a $\sin\theta$ as θ and a are small,

$m_2 = m_1 x / y$

i.e. $m_2 = 1 (900/300) = 3$ kg

to account for a $\sin\theta$, m_2 is taken to be 3.05 kg

Now using equation 2 the value of θ was calculated,

$\theta = \sin^{-1} (h / a)([(m_2 y - m_1 x)/ m_1 x])$

substituting the values we find

70

$\theta = 4.48°$

i.e. $\theta \approx 5°$

Verifying equation 3,

$Pw/g \geq (m_1\ x^2 + m_2\ y^2)\ \sin2\theta\ /\ [x\ (h - x\sin2\theta\ )]$

LHS = 2

RHS = $[1\ (1.2)^2 + 3.05\ (0.3)^2]\ \sin10\ /\ 1.2\ (\ 1-\ 1.2\ \sin10\ )$

= 0.313

As, LHS > RHS the system will rotate freely till $\Phi = 180°$

Hence the designed parameters were as follows

$m_1 = 1$ kg

$m_2 = 3.05$ kg

$\theta = 5°$.

$x = 1200$ mm

$y = 300$ mm

$a = 150$ mm

$h = 934$ mm

4. FABRICATION AND IMPLEMENTATION

The REVOLVING MATERIAL TRANSFER SYSTEM was fabricated in the central maintenance of BAL.

The components/parts used in above designed were fabricated as follows:

- TRAY :

    Material used – 3mm steel sheet

    Dimensions of the tray 26 X 18 X 5 cm

    QTY - 1

    The steel sheet of 36 X 28 was cut on shearing m/c then squares of 5 X 5 cm was cut out from the corners on nibbling m/c. the tray was then made out of the cut shape by bending it on bending m/c. the corners were spot welded.

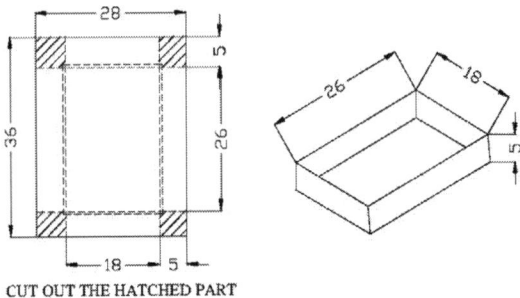

CUT OUT THE HATCHED PART

Figure 39: Tray

- TELESCOPIC ARM:

    Material used – GI Pipes

    Dimensions (i) Ø30 o.d, Ø 27 i.d, 600mm length

    (ii) Ø 25 o.d, Ø 23 i.d, 900mm length

Through holes of Ø10 were drilled at a pitch centre distance of 100 mm on both the pipes.

The inner pipe was inserted inside the outer and bolts were used to fix a total arm length of 1.2 m.

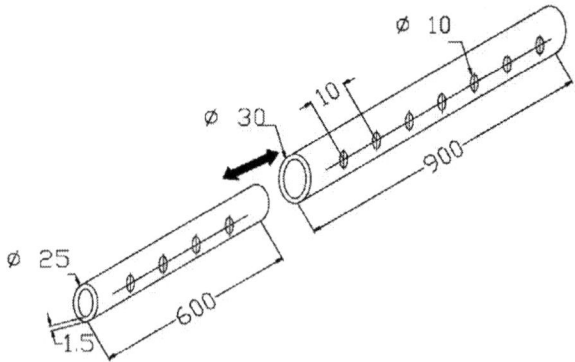

Figure 40: Telescopic Tray Arm

- BEARING HOUSING

    The bearing housing consisted of three parts

    (i)   the casing

    (ii)   the bearing

    (iii) the bearing bush

72

(iv)   the bearing axle

The casing was made out of steel sheet 4mm thick of the shown dimensions.

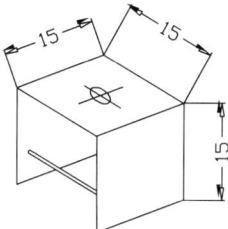

Figure 41: Bearing Housing

A deep groove ball bearing was used having o.d Ø52, i.d Ø20

The bearing bush was manufactured of mild steel as per the shown drawing.

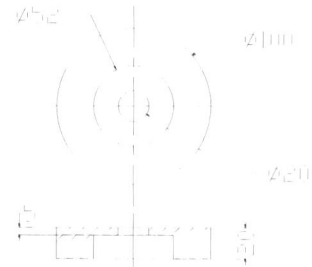

Figure 42: Bearing Bush

The bearing axle was manufactured out of case hardened alloy steel as per the shown drawing.

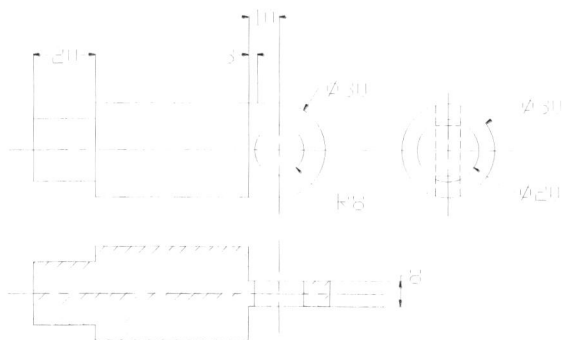

Figure 43: Axle

73

- FOUNDATION COLUMN

  Materials and dimensions:

  1.   Standard Square channel 80 x 80 mm having 850 mm height.

  2.   MS plate 5mm thick of 200 x 200 mm size

  The MS plate with holes of Ø16 mm at the corners was welded at the bottom of the square channels as shown.

- COUNTERWEIGHT and ARM

  A plastic bin of size 18 x 15 x 10 cm was bolted on two L channels which were welded on the casing of the bearing housing. Suitable weights could be added to the bin as counterweight.

  The bearing axle inclination was kept adjustable by providing two bolts for adjustment of the suitable angle as shown.

- STOPPERS FOR WORKING ANGLE ADJUSTMENT

  Two L shaped stopper channels were mounted on the axle from the top as shown. The angles could be easily adjusted by loosening the nut from the top and re-tightening after the required working angle was fixed.

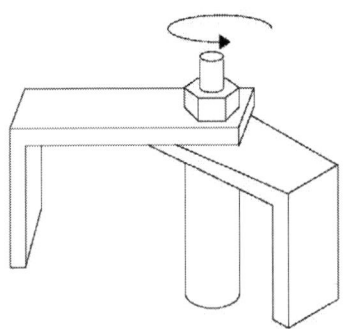

Figure 44: Stopper

After fabrication of all the components required, they were assembled. Several trials were conducted and the equipment was found to perform consistently and satisfactorily.

Figure 45: Photos of revolving Tray During trials

5. ADVANTAGES

- Low fabrication, running and maintenance cost.
- Reduces operators handling thus reduces operators fatigue.
- Robust, simple, compact and flexible design makes it suitable for almost all short distance material transfer applications
- Consumes no electrical power as works on counterweight principle.
- The effective productive time increases as the operator can utilize the time for loading/unloading jobs from m/c which was otherwise wasted in material transfer.
- If intelligently mounted on line, it does not block the passages for operator's movement

6. LIMITATIONS

- Application is restricted to small distance material transfer
- The revolving long arms demand more space, and may obstruct the operator's movement.
- Not suitable for heavy components.

FUTURE SCOPE

Endless efforts can be made to continuously improve upon the material handling systems. The use of KARAKURI concept would cut upon the cost and labour involved in material handling to a great extent.

It is possible to completely replace the powered conveyors by innovative thinking and utilizing gravity. The concept lets the door open for entire low cost automation of the line.

KARAKURI KAIZEN meaning "Good Changes with KARAKURI" makes an effort in designing and implementing some very simple, mechanically operated equipment to improve material handling.

The project shows that the equipments can be easily implemented parallely to all similar application areas. The use of such material handling systems is possible in almost all small to large scale production industries.

One may easily predict that KARAKURI would prove out to be a revolutionary concept for industrial development.

References:

1. L.C. Jhamb; Industrial Management II; Everest Publishing House 4th edition;

2. Shigley; Theory of Machines; TATA McGraw Hills Publications; 8th edition;

3. FESTO Product Overview 2005;

4. www.worldofkarakuri.com

5. John Smith; Research Paper on "Low Cost Automation"; Stanford University, USA;

List of Figures:

Made in the USA
Middletown, DE
02 February 2018